# FEATURES

SUMMER 2024 • NUMBER 40

C000113072

## Plough

# INSIGHTS

# ARTS & LETTERS

# DEPARTMENTS

# WEB EXCLUSIVES

Read these articles at *plough.com/web40*

# Plough

## ANOTHER LIFE IS POSSIBLE

**EDITOR:** Peter Mommsen

**SENIOR EDITORS:** Shana Goodwin, Maria Hine,
Maureen Swinger, Sam Hine, Susannah Black Roberts

**EDITOR-AT-LARGE:** Caitrin Keiper

**BOOKS AND CULTURE EDITOR:** Joy Marie Clarkson

**POETRY EDITOR:** A M Juster

**ASSOCIATE EDITORS:** Alan Koppschall, Madoc Cairns

**CONTRIBUTING EDITORS:** Leah Libresco Sargeant,
Brandon McGinley, Jake Meador, Santiago Ramos

**UK EDITION:** Ian Barth

**GERMAN EDITION:** Katharina Thonhauser

**COPY EDITORS:** Wilma Mommsen, Priscilla Jensen, Cameron Coombe

**DESIGNERS:** Rosalind Stevenson, Miriam Burleson

**CREATIVE DIRECTOR:** Clare Stober

**FACT CHECKER:** Suzanne Quinta

**MARKETING DIRECTOR:** Tim O'Connell

**FOUNDING EDITOR:** Eberhard Arnold (1883–1935)

Plough Quarterly No. 40: The Good of Tech
Published by Plough Publishing House, ISBN 978-1-63608-148-9
Copyright © 2024 by Plough Publishing House. All rights reserved.

**EDITORIAL OFFICE**

United Kingdom
Brightling Road
Robetsbridge
TN32 5DR
T: +44(0)1580.883.344

North America
151 Bowne |Drive
Walden NY 12586
T:845.572.3455
info@plough.com

**SUBSCRIBER SERVICES**

Unit 6, The Enterprise Centre
Kelvin Lane, Crawley RH10 9PE
Tel: 0800.018.0799
Plough@subscription helpline.co.uk

Australia
4188 Gwydir Highway
Elsmore, NSW
2360 Australia
T:+61(0)2.6723.2213

Plough Quarterly (ISSN2372-2584) is published quarterly by
Plough Publishing House, PO Box 398, Walden, NY 12586.

Individual subscription £24/ €28 / $36 per year.
Subscribers outside of the United States and Canada pay in British pounds or euros.

Front cover: Pilar López Báez, *The Good of Tech*, mixed media, 2024.
Inside front cover: Yalim Vural, iPhone photography. Used by permission.
Back cover: Leonardo Ulian, *Technological Mandala 92 – Model of Everything*, 2016.
Used by permission.

## ABOUT THE COVER

One instance of good tech is the
Dexcom system for managing
diabetes, illustrated here by Spanish
artist Pilar López Báez and discussed
in more detail by Peter Mommsen in
his editorial on page 10.

# LETTERS

Readers respond to *Plough*'s Spring 2024 issue, *The Riddle of Nature*. Send letters to *letters@plough.com*.

## THE RIDDLE OF NATURE

Nature is indeed a riddle. Nearly every human has experienced the healing powers of nature. Time spent outdoors can be deeply restorative and is known to help maintain mental health. At the same time, nature can be ruthless and deadly.

The truth is that nature does not care for you at all. Nature will just as quickly kill you and give your energy away to other organisms as it will heal you and make you feel wonderful. Nature is completely agnostic about whether or not you survive. We must recognise that nature is ruthless and breathtakingly beautiful, all at the same time. Just like all of us humans.

*Nathan W Ferrell, Falmouth, Maine*

## UNDERSTANDING NATURE'S BOOK

*On Peter Mommsen's 'The Sadness of the Creatures':* This article is brilliant. Much of what we read these days is junk. Folks think they have something to say but most is recycled stuff. Well, Whitehead did say Western philosophy is mostly comments on Plato. So there is little new under the sun, so to speak. But I think your article for me was new. I like that you are not afraid to cast a wide net and bring in Auden, the Nazis, Anthony in the desert, Woolston, Augustine, Hut, etc. I see humanity rushing to a precipice and instead of taking warning, adding fuel. Only the humanities, the study of religion, the arts, history, etc., will save humanity over the next several hundred years. Of course, humans will need the life skills of agriculture and animal husbandry, carpentry, analogue repair, blacksmiths. Much of it has been lost but, like the words of the old monks in the monasteries, it will be refound and re-energised. Your communities are a guiding light.

*Bob Kambic, Baltimore, Maryland USA*

## GOLDEN PESTS

*On Clare Coffey's 'Dandelions: An Apology':* The dandelion has always been my favourite flower, precisely for its 'here I am again, like it or not!' persona. There are few pictures more beautiful than a meadow of gold and few things more delightfully surprising than a brave little golden head pushing up through a crack in the cement pavement. We cultivate vegetables for a living, so technically the dandelion should be a pest. Yet each time we are on hands and knees weeding the greenhouse and one of my boys finds one of these sweet pests, it is tenderly plucked and brought to me, 'Here, Mom, for your hair'.

*Sonya Woolston, Ulster Park, New York USA*

## SEEING EVERYTHING

*On Daniel J D Stulac's 'Promised Land':* Years ago, there was a TV show, *Corner Gas*, that took place on the Canadian prairies. Two of the residents are sitting on a car looking out over the prairie. A nonresident comes by and says, 'I can't see anything!' One of the residents replies, 'No, I can see everything.' Thanks for your 'I can see everything' perspective.

*Nicolai Hansen, Rockledge, Florida USA*

You have so eloquently captured the journey from adult (jaded?) comparisons – 'desolate expanse' – to childlike acceptance and wonder – 'the sky . . . it's alive!' Mountain grandeur gets more promotional leverage than plains plainness. Both have a starkness. But if one makes the choice to actually engage and not virtually escape life on the prairies, then one has the opportunity, over time, to have one's lust for the spectacular purged by the subtle and the nuanced.

*Doug Reichel, Moose Jaw, Saskatchewan Canasa*

## OUR PREDATOR SPECIES

*On Tim Maendel's 'Why We Hunt':* Amid all the philosophising and theologising it is refreshing to read of someone who actually interacts with nature at the visceral level of seeking the meat he consumes by his own hand while reflecting on what that says and means. I too am a hunter, in particular a bow hunter and a bird hunter behind my beloved field-bred English Springer Spaniels. I too reflect on what it means to be another of the apex predators, and on the precious value of my prey, and on how I and they fit into the greater scheme of God's creation.

*Herb Evert, Cottage Grove, Wisconsin USA*

Maendel's reflection on hunting deer and turkey is sure to evoke squeamish reactions in some. I share the squeamishness. Living in the country, with animals, I've seen several situations where mercy killing was clearly indicated. Sometimes I've been able to do the job, sometimes not. I met a young doe once who lurched towards me rather than bounding away as a healthy deer would have done. I'd read about chronic wasting disease in deer and was pretty sure I was looking at it. I stomped and yelled and scared her away. If I'd been carrying anything that could be used to euthanise a deer, which I wasn't, I'm still not sure I could have killed a deer.

It is squeamishness about killing animals that reminds us of ourselves, though. I know that because I am capable of taking lives that look just a little more alien. If you want beans, you kill beetles; the fool's way is to spray poisons that kill helpful animals and produce allergies; the better way is to scoop beetles into a pill bottle and then, when the bottle is full or the beans are bee-tle-free, add water or alcohol. If you want fruit trees, you kill the vines that drag them down. If you want a pretty rosebush to look at, instead of a tangle of briars across the path, you kill the branches that try to grow in the wrong place and direction.

How much consciousness a bean beetle, a dog tick or a honeysuckle vine may have, I'm not sure, but I am sure that with all the consciousness they have they prefer to stay alive. I kill them.

Most of us can afford not to share what Tim Maendel makes sound like a *vocation* to hunting. As a species, though, we need people like him. In the future, as in the past, we may come to depend on those people.

*Priscilla King, Gate City, Virginia USA*

## MIRACLES OF NATURE

*On Norann Voll's 'Lambing Season':* I live in the Basque country, Spain, as my wife is Basque. I was a shepherd for four years in England, after which I became a 'lecturer in agriculture' for 28 years before retiring here. In my work as lecturer, I used to teach students various practical sessions such as foot-trimming cows, reversing trailers and of course lambing.

I well remember one lambing session. I told the students to observe one particular ewe, who didn't need any assistance in producing one healthy, strong lamb. As we had plenty of time, I said to them, 'Right, let's wait and see how long it takes that newborn lamb to get up and find some milk.'

Eventually we saw several attempts of the lamb butting something, rather clumsily, around the ewe's udder. When we saw its head under the udder and its tail wagging, we knew it had hit bullseye.

'How long?' I asked my students.

'Forty minutes,' they replied.

'So,' I asked them, 'How did the lamb know where to go?'

'Instinct,' they replied.

'Yes, OK, but what is instinct? Where did that lamb get it? Who gave them that instinct? Surely only the Creator could have,' I said.

'Sir, you don't believe in God, do you?' they enquired, quite seriously.

'Why not? Look at the miracle you've just observed.'

*Paul Attard, Elorrio, Bizkaia, Spain*

## A LACK OF IMAGINATION

*On Marianne Wright's review of Grace Hamman's* Jesus through Medieval Eyes*:* I think a core problem with the modern world is its lack of imagination. This isn't universally true, of course, but we put a heavy emphasis on science, maths, engineering and such and have become dismissive of the humanities where we develop the imagination and learn to question the facts the maths and sciences point to. But just as everything from the Middle Ages should not be preserved, so also not everything we can use maths and science to accomplish ought to be accomplished. We are dealing with some of those things now. It doesn't mean we should abandon things like AI, but we should be much more thoughtful than we are about how we use them and what can be done with them. Of course, we cannot help ourselves; we want to play with these sorts of things. If we did more than just think about what we might do but also imagined what the results might be, we might have fewer problems.

*John D Wilson, Jnr, Centerville, Massachusetts USA*

## Loving the University

*At Washington University in St Louis, Missouri USA, an uncommon community is growing.*

**John Inazu**

One of the central motivations of my academic career has been to ask how Christians can flourish in nonreligious colleges and universities. I would like to think that Christians can actually model some of what the university purports to be. Five years ago, I founded The Carver Project, a staff-based ministry at Washington University in St Louis, Missouri. In our efforts to serve and connect university, church, and society, we have also furthered the university's goals of interdisciplinarity, student engagement, and town-gown relations.

Consider first interdisciplinarity. It's not easy to incentivise work across disciplinary boundaries – schools and departments are often geographically isolated, teaching schedules and departmental norms vary, and staff are busy enough trying to maintain relationships with their own colleagues. The Carver Project's interdisciplinarity is rooted in friendship. We learn about each other's gifts and passions over shared meals and conversations. And collaborations grow out of these relationships, like the course on 'Law, Race and Design' I teach with graphic designer Penina Laker or the class on 'Markets and Morality' taught by Peter Boumgarden in our business school and Abram Van Engen in our English department. Other collaborations such as our panel discussions or public events bring together unusual groups of staff from across different disciplines.

We further student engagement through reading groups that often meet in staff members' homes. Every month, a group of law students gathers at my house for food, drink and a book discussion related to Christianity and law. In the classroom, I teach students about the law. At my house, I teach them why it matters. It's not just me – Carver Project staff host similar groups in art, medicine, business, English and other subjects. My colleagues and I

---

*John Inazu is the Sally D Danforth Distinguished Professor of Law and Religion at Washington University in St Louis and author of* Learning to Disagree: The Surprising Path to Navigating Differences with Empathy and Respect *(Zondervan, 2024).*

Carver Project law students' reading group.

seek to model a kind of uncommon community that engages students more holistically than is possible in a secular classroom.

Finally, The Carver Project helps lower the proverbial wall between the campus and its surrounding communities. Our staff fellows are fully immersed in both the university and their local church. We care deeply about both institutions and we sense the ways they can partner more effectively to further the flourishing of the surrounding community. We lower relational barriers to partnership every time our members speak at local churches or welcome our church friends to campus events.

The Carver Project is not on a secret mission to take over Washington University. Rather, we seek to be a faithful presence in the university that partners with it to advance common efforts. Through this work, my colleagues and I have gleaned three key insights in establishing a community of Christian staff within a non-Christian university: 1) we began by learning the university's language and culture, 2) we built things together and 3) we took risks, knowing that our academic identities are anchored in a far greater story than that of the university.

**Learning language and culture:** Years ago, I volunteered with the Christian ministry Young Life. One of Young Life's mantras is 'earn the right to be heard'. Our staff have spent their academic careers earning the right to be heard. They are some of the nation's leading experts in their fields, they are caring and effective teachers and they serve the university in seen and unseen ways. We have flourished in these ways because we took the time to learn the university's language and culture – its spoken and unspoken norms, its peculiarities, its demands and its weaknesses. Rather than respond with fear or anxiety to unfamiliar or even off-putting rules and cultures, we have learned how to thrive within them. That doesn't make our experiences free of pain or disappointment. But it does mean we can thrive within an environment that does not always align with our values or comfort levels.

**Building together:** From the beginning of The Carver Project, I told my colleagues that I was only willing to lead the organisation if others would join with me to help steward it. Two years ago, my friend and colleague Abram Van Engen succeeded me as The Carver Project's executive director. He's not leading how I led and he sometimes makes decisions that I would not make. But that's the point. The best way to avoid a vanity project is to disperse the leadership, vision and control. I am excited to be part of an organisation that Abram leads and I am excited to keep doing great work together.

**Taking greater risks:** University staff have many different roles and responsibilities, but many of us benefit from an abundance of resources and the protections of tenure. In building The Carver Project, we endeavoured to take greater risks with our time, money and reputations. The past five years have been much harder than I anticipated – for me, risks have turned into costs, particularly when it comes to time. Nor am I the only one who has sacrificed. I think of Allie and Kelly, the two law students who built the organisation with me from the ground up. I think of the students who followed them, our faculty fellows who built programmes and my colleague Abram who stepped into the leadership role. I think of our managing director, Shelley and a long line of other staff, board members and donors. Every one of them took risks and made sacrifices.

There is a chance that we have collectively built something that will far outlast us. It's still just a chance – fraught with the messiness of people, personalities and contingencies. Part of taking risks is not always knowing how your work will be received or even if you will be around to know. But even in these fits and starts that have marked our first five years together, we have glimpsed the possibility of something far greater than we could have asked or imagined.

I hope that more Christians will come to see the good in non-Christian higher education instead of fixating on its dangers and imperfections. These institutions will never feel like home: the days of Christian mottos centring America's most elite colleges are over; the baccalaureate ceremony is a dying relic at most institutions; and in most cases, the most elite schools can no longer name their purpose at all. They are rich, powerful and for the most part, listless. But when Christians show up to be part of them – by learning the language and culture, by partnering together in the work at hand and by taking risks – the Lord can do amazing things. ⤜

## Locals Know Best

*Inspired by the Psalms, Loom International supports local communities in the developing world.*

### Janna Moats

Shortly after our 25th wedding anniversary, my husband died, aged 46. He had been fighting cancer for several years. I was left with my two young adult children and no idea where to go next. This wasn't the life I had imagined 25 years before on our wedding day. I decided to take a course about the needs of children in low-income countries. I learned about the many challenges these children face: exploitation, malnutrition, abuse, neglect and lack of education. I knew I wanted to help vulnerable and suffering children.

In Psalm 10, the psalmist asks, 'Why, O Lord, do you stand far off in times of trouble?' Like me, the psalmist had assessed the world around him and had seen that the wicked exploited the weak. Like me, he believed that God is good and he wanted to know why a good God did not get involved. In verse 14, the answer comes: 'But you, God, see the trouble of the afflicted; you consider their grief and take it in hand. The helpless commit themselves to you; you are the helper of the fatherless'.

This passage inspired me to found Loom International. In 2007, my vision took off. I started meeting deeply dedicated people in India, Bangladesh, Romania, Mexico and East Africa who were founding schools, building foster care homes for children whose parents had died from AIDS, and starting clinics and after-school programmes. They had often begun with minimal financial support and guidance, but they saw a need and started doing something. They knew their people and their community. Yet often the projects were struggling to keep afloat, even though their founders were passionate and committed. When resources and funding are scarce, staff can become burnt out and discouraged. Loom began by listening to these people and asking what they needed in order to continue and expand their mission. Where they needed help, we backed them up.

In time, Loom became a support network for local communities, and in particular, the vulnerable children within those communities. We believe that children are intended to grow and mature in families and within communities committed to their flourishing, and that if you help the smallest, weakest and poorest in a society to thrive, that society will flourish. ➤

*Janna Moats is the founder of Loom International and continues to serve as director of learning resources. She currently lives in Portland, Oregon. Loominternational.org*

## Winners of the Fourth Annual Rhina Espaillat Poetry Award

Congratulations to winner Jennifer Fair Stewart for her poem 'Blackberry Hush in Memory Lane' and to finalists Michael Manning for 'A Lindisfarne Cross' and Laura R Eckman for 'Fingered Forgiveness', all published in the pages of this issue. The winners were announced at a livestreamed event with Rhina P Espaillat. The award is for a poem of not more than fifty lines that reflects Espaillat's lyricism, empathy and ability to find grace in everyday events of life. The 2024 competition attracted over 800 poems.

**Jennifer Fair Stewart** grew up running wild across midwestern acres, sojourned in cities and now runs semi-respectably through southwestern desert suburbs. She has been working as a teacher for over twenty years, in addition to stints as a sales clerk, house cleaner, library assistant, caretaker, secretary, and retreat speaker. As a child, she won her school's annual competition for the Young Authors Conference twice and has been writing off and on ever since. Her poetry has been published or is forthcoming in *Heart of Flesh, The Orchards, Quiet Diamonds, Crescendo,* and *Trampoline.* Her chapbook *Marginalia: An Interactive Book of Hours* is forthcoming in 2024 with Orchard Street Press. Her poem 'Blackberry Hush in Memory Lane' appears on page 33.

**Michael Manning** lives on the Isle of Man and has spent a lot of time with people who are homeless. He has written for *ISAAC (International Substance Abuse and Addiction Coalition),* the *Baptist Times,* and *EthicsDaily.com.* He is the author of the

A church service under a tree in Engikaret, Tanzania.

nonfiction *No King but God* (Resource Publications, 2015) and is the Eighth Manx Bard. His poetry has been published in *Manx Reflections, the Manx Independent, Dreich Planet#2,* and *The 2025 URC Prayer Handbook: Immersive Joy* (forthcoming). He now works as a healthcare assistant. His poem 'A Lindisfarne Cross' appears on page 65.

**Laura R Eckman** was born and raised in a small town just north of Pittsburgh, Pennsylvania. A multilingual poet, she has written and published in both English and BCS (Bosnian-Croatian-Serbian). She currently lives in southern Europe, where she works for a foundation that offers language classes and seminars, and also promotes the arts. Her poem 'Fingered Forgiveness' appears on page 102.

*Plough's* 2025 poetry competition is now open. The winner receives $2000, and two finalists receive $250. All three will be published in *Plough*. Submit your new poems at *plough.com/poetryaward.*

## A Word of Appreciation

After four years as *Plough's* poetry editor, A M Juster has let us know that with this issue he is concluding his service in that role. As his fellow editors, we're immensely grateful for the lasting mark Juster has left on *Plough's* poetry programme, and wish him the best for his many other endeavours. His successor will be announced in the coming months.

A poet, translator, and essayist, Juster co-founded *Plough's* Rhina Espaillat Poetry Award in 2021. Each year he evaluated hundreds of submissions singlehandedly to select the finalists, with the winners selected by the full editorial board for publication in the subsequent issue. In addition, for each other issue of the magazine he invited an accomplished poet to submit two or three original poems.

Juster is known as a key figure in the New Formalist school, that aims to create poetry that speaks to a wide audience through the use of classical forms and everyday themes. He's also a champion of light verse, to which he's made his own memorable contributions in the 2015 collection *Sleaze & Slander*. All the same, as an editor, Juster has been ecumenical in his commissions, with an appreciation for a wide variety of schools and styles.

In an interview for *Plough* with Australian poet Stephen Edgar, Juster said: 'If you look at the people whose poetry I admire, they're going out and doing things that they are passionate about. And I think that ultimately flavours their poetry. I liked having challenges out in the world. I think that gave me more interesting poems. When I was younger, sometimes I would just sit down and stare at a blank page, but now not so much. I need to have something that I'm thinking about that grips me to go and really want to spend the time in this struggle.'

I know many readers will join me in thanking A M Juster for his dedication to bringing great poetry to the pages of *Plough*. We'll miss his enthusiasm, wisdom, and wit, but look forward to continuing to follow his work at amjuster.net or @amjuster. Any readers with even just a budding interest in poetry will do themselves a favour if they pick up his marvellous 2021 collection *Wonder and Wrath*.

—*Peter Mommsen*

## About Us

*Plough* is published by the Bruderhof, an international community of families and singles seeking to follow Jesus together. Members of the Bruderhof are committed to a way of radical discipleship in the spirit of the Sermon on the Mount. Inspired by the first church in Jerusalem (Acts 2 and 4), they renounce private property and share everything in common in a life of nonviolence, justice, and service to neighbours near and far. There are 29 Bruderhof settlements in both rural and urban locations in the United States, England, Germany, Australia, Paraguay, South Korea, and Austria, with around 3000 people in all. To learn more or arrange a visit, see the community's website at *bruderhof.com*.

*Plough* features original stories, ideas, and culture to inspire faith and action. Starting from the conviction that the teachings and example of Jesus can transform and renew our world, we aim to apply them to all aspects of life, seeking common ground with all people of goodwill regardless of creed. The goal of *Plough* is to build a living network of readers, contributors, and practitioners so that, as we read in Hebrews, we may 'spur one another on towards love and good deeds'.

*Plough* includes contributions that we believe are worthy of our readers' consideration, whether or not we fully agree with them. Views expressed by contributors are their own and do not necessarily reflect the editorial position of *Plough* or of the Bruderhof communities.

# The Artificial Pancreas

*How can we live well with technology?*

**PETER MOMMSEN**

**W**HEN I TOOK MY SON and a bunch of his cousins swimming at the lake one afternoon last summer, I didn't plan to lose a piece of someone's pancreas. The boys were practicing flips off the dock. After an hour, their daring grew to the point that I called time. As they straggled onto the beach, I noticed that my nephew, whom I'll call Tristan, was missing his smartphone, that he'd stuffed into a waterproof

case. While the loss would be bad news for anyone, for Tristan it was far worse: the device is part of the management system for his type 1 diabetes, a condition in which the pancreas stops producing insulin. Until the 1920s, when scientists discovered how to extract insulin from dogs, type 1 diabetes was fatal in the short term. Even for decades afterwards, a diagnosis meant a regimented life with daily glucose tests, exactly

Pilar López Báez, *The Good of Tech*, mixed media, 2024 (detail).

timed insulin injections and a harshly restricted diet; despite all that, the disease still drastically reduced life expectancy and came with a high risk of side effects including blindness, heart and kidney failure and limb loss.

Today, technology enables Tristan to live much like any other energetic twelve-year-old. Starting in 1978, scientists developed more sophisticated forms of insulin (synthetic, no dogs needed). In 1999, medical technology companies started producing continuous glucose monitoring systems (CGMs) that could track blood sugars using a probe placed under the skin. Then in 2017, easily portable CGMs became widely available. These give real-time readings on blood glucose levels. Today's CGM talks via Bluetooth to a pump that delivers calibrated doses of insulin via a cannula. A cellphone controls the system and provides data to the user. This multi-component system is sometimes called an 'artificial pancreas' for its ability to mimic the natural organ's function.

It's not the same as a natural pancreas, of course – the technology is glitchy and the human body is endlessly variable, so people with type 1 diabetes still struggle to maintain healthy glucose levels, knowing that high levels are unhealthy and lows can be life-threatening. Manual inputs are needed for every carbohydrate eaten and both the sensor and pump need changing out every few days, making those affected not only dependent for their very lives on insulin but also, for good management, on the manufacturers of the smartphone, pump, CGM and various other consumables. (That's aside from the weird precarity and expense of insulin supply in the United States, where people with the condition regularly beg strangers for emergency doses on Reddit or die from attempts to ration their supply.) All the same, for someone with type 1 diabetes, this technology promises the chance at a greatly, if imperfectly, liberated life.

That is, until a piece of the system sinks to the bottom of a lake. From a check of Tristan's wristband, it looked as if a cheap plastic clip had popped open when he swam to shore. This was a big deal: I knew from his parents that the system was new, having been finally approved by their health insurer, which, in addition, doesn't cover the smartphone. For half an hour, we dived to search for the thing, but at a three-mitre depth, the water was too murky to see much. All that remained was to hike home to confess to his parents, unsure what the loss meant. This time, luck was with us: they hadn't thrown out the replaced older model yet. The artificial pancreas was soon working again and we celebrated with glucose-heavy ice cream.

---

## We seem to be on the cusp of a technological transition that may, once again, revolutionise our lives for good or ill.

---

This is a happy story about the good of technology, the kind that can transform a kid's life. It's a technophile parable in which an artificial system that combines mobile software and automated hardware literally plugs into a human body, helping it to thrive in a way it wouldn't if left to nature. The fact that such a technology exists in the first place is a testimony to the real achievements of Silicon Valley and its 'Californian ideology', which merged freewheeling innovation and entrepreneurial capitalism in a mission to change the world.

In many quarters today, the promises of the Californian ideology have come to seem suspect. Contrary to the hopes of its early enthusiasts, the internet didn't bring all of humanity together in the 1990s; neither did smartphone-enabled social media in the 2000s. In Washington, London and Brussels, political leaders have long since stopped associating tech with utopian visions of global

harmony, instead blaming it for distraction, polarisation, addictions to porn and gambling, the trivialisation of culture, loss of privacy and work-life balance, and fears that automation may push millions out of a job.

Growing scepticism has done little to stop the acceleration of change. Advances in artificial intelligence seem poised to bring us to the next technological watershed – very soon. That's at least what AI's cheerleaders promise and its critics fear. 'Generative' artificial general intelligence based on large language models – for example, OpenAI's ChatGPT, Anthropic's Claude and Meta's Llama – are rapidly increasing in power from version to version. Industry leaders such as Anthropic's Dario Amodei believe that well-developed 'agentic' AI may be just around the corner: that is, AI that can take actions in the real world, such as booking a vacation or planning a wedding.

Meanwhile, specialised versions of AI are transforming any number of industries. In pharmaceuticals, an AI breakthrough this year has solved a 50-year-old problem of molecular biology by enabling scientists to quickly model the shape of proteins in three dimensions; other AI applications promise greatly accelerated drug development and testing. In healthcare, AI algorithms can detect cancers at early stages with high accuracy, often outperforming experienced radiologists. In retail, AI helps manage inventory and predict demand patterns, improving the efficiency of supply chains. And in the automotive industry, while fully autonomous vehicles still remain a future goal, AI systems in cars are already improving driver safety.

Other developments are more unsettling. Several nations are already fine-tuning autonomous weapons systems – that is, killer robots – for use on the battlefield. In human genetic engineering, AI is accelerating genomics research and so bringing forward the day when it will be possible to create designer babies. And then, of course, there's the apocalypse: some AI experts worry that the technology could become powerful enough to cause human extinction and the Pentagon seems to be taking seriously the risk that AI might somehow trigger nuclear war.

We seem to be on the cusp of a transition that may, once again, revolutionise our lives for good or ill. It's a good time to ask how we should respond.

I N 1992, JUST AS the Californian ideology was picking up steam, Neil Postman published *Technopoly,* a classic work of cultural criticism that explores the relationship between humans and our tools. The book opens by quoting Plato's account of the myth of Thamus, an Egyptian Pharaoh. The king is approached by the god Theuth, who presents various inventions, including writing. Theuth suggests that writing will improve both the wisdom and memory of the Egyptians. Thamus, however, objects:

> This invention will produce forgetfulness in the minds of those who learn to use it, because they will not practise their memory. Their trust in writing, produced by external characters that are no part of themselves, will discourage the use of their own memory within them. You have invented an elixir not of memory, but of reminding; and you offer your pupils the appearance of wisdom, not true wisdom, for they will read many things without instruction and will therefore seem to know many things, when they are for the most part ignorant and hard to get along with, since they are not wise, but only appear wise.

Reflecting on the myth, Postman observes that 'every technology is both a burden and a blessing; not either-or, but this-and-that'. The very tool that promises to expand humans' knowledge also robs them of their oral tradition. It makes new winners – adepts of the written word – and new losers: the elders who had been memory's guardians. To update Postman's telling, Theuth was setting in motion a chain of discoveries that would, in due course, transform words and images to digital tokens and so give rise to ChatGPT.

Postman argues that digital technology, that reduces all meaning to symbols that a machine can compute, tends to change our view of nature and ourselves. We come to regard nature as mere information to be processed and human beings as mere processors of that information. To the extent we allow the machine's logic to become our own, we'll become 'tools of our tools', in Thoreau's phrase.

The real-life Postman practised what he preached, eschewing digital tools such as word processors and cruise control. He chose to remain in an analogue world that was vanishing as he wrote. In surveying the technological revolutions of the past, he's always glancing wistfully backwards. (This sometimes gives him blind spots; for example, he makes much of the flaws of modern medicine, while giving short

Leonardo Ulian, *Technological Mandala 83*, electronic components, copper wire, and speakers on paper, 2015 (detail opposite).

shrift to its achievements.) He's fascinated by the unpredictability of the effects of innovation, noting how the invention of the printing press by the pious Johannes Gutenberg had the unintended consequence of destroying the unity of Western Christendom. With each such shift, a world ends:

> Technological change is neither additive nor subtractive. It is ecological. I mean 'ecological' in the same sense as the word is used by environmental scientists. One significant change generates total change ... A new technology does not add or subtract something. It changes everything.

Just as introducing or eliminating one species can change a forest ecosystem, so a new technology can remake the whole fabric of our lives.

THAT'S THE KIND of ecological change that Jonathan Haidt chronicles in his new book, *The Anxious Generation: How the Great Rewiring of Childhood Is Causing an Epidemic of Mental Illness.* A social psychologist at New York University, Haidt compiles evidence of a surge in an array of ills that began around 2010, just when smartphones and Like-button-powered social media became part of childhood and adolescence. Over that period, he finds a spike among US teens in reported rates of anxiety, depression and loneliness. Since 2010, suicide rates have risen 91 percent for American boys and 167 percent for girls, while emergency room visits for self-harm have risen 48 percent for boys and 188 percent for girls.

Provocatively, Haidt argues that this 'surge of suffering' not only coincides with the advent of smartphones, but results from it. According to his findings, constant access to social media exacerbates girls' image-consciousness and vulnerabilities, while internet-connected devices enable boys to retreat into virtual experiences to escape the challenges of the real world.

Some critics dispute Haidt's interpretation of these trends. But practically, as a parent with three teenagers, I don't have years to wait for scientific certainty, nor do other parents. We're left with weighing up what evidence there is and trusting our own eyes. It doesn't take big data sets to see how rapidly and deeply childhood and adolescence have been transformed over the past decade and a half. But it does take a considerable amount of go-against-the-flow to act on what we see.

The impacts aren't limited to mental illness rates. Consider the sheer amount of time kids spend on electronic media. In the early 1990s, US teens spent a little less than three hours a day looking at screens, mostly televisions. Today that number has gone up to six to eight hours a day, even more for those from low-income households (that's just counting leisure time, not school-related use). There's little mystery as to why. Multinational tech giants, armed with algorithms and vast amounts of data, have become ever better at keeping adolescents engaged with their products. That's lucrative for them, but comes at a massive opportunity cost to kids, who are enticed to surrender vast chunks of their teenage days and years to screens.

It's not enough, then, just to ask what teens are doing online and whether some of it might be harmful. An equally important question is: what are all the things they *aren't* doing offline? Drawing on Allie Conti's reporting, Haidt tells the story of Luca, a young man from North Carolina:

> Luca suffered from anxiety in middle school. His mother withdrew him when he was 12 and allowed him to study online from his bedroom. Boys of past generations who retreated to their bedrooms would have been confronted by boredom and almost unimaginable loneliness – conditions that would compel most homebound adolescents to change their ways or find help. Luca, however, found an online world just vivid enough to keep his mind

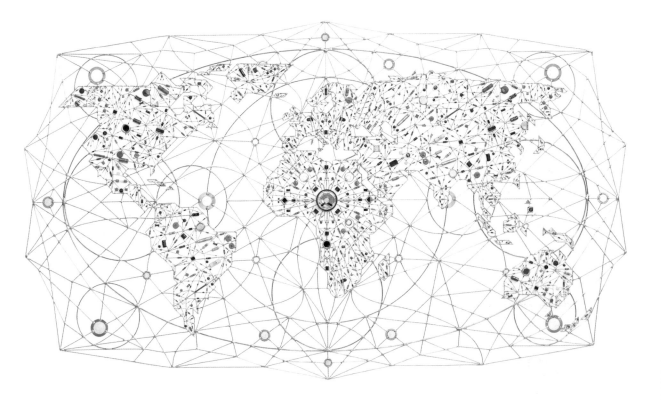

from starving. Ten years later, he still plays video games and surfs the web all night. He sleeps all day.

Luca's story is an extreme version of behaviour that most parents of adolescent boys will recognise. As Postman would point out, it's a playing-out of a dynamic that's built into the technology, one that grows more powerful as the online experience becomes more immersive. Unless counterbalanced by offline interests, work and friendships, the virtual world can swallow up a life.

Luca seems to be far from alone. There are substantial online communities for NEETs ('not in education, employment, or training') and *hikikomori,* a Japanese term for young people who spend their lives as digital hermits in their bedrooms. One Reddit user explained when asked why he became a NEET: 'Growing up I had zero friends and would just skip school to stay at home and play video games. Nothing much has changed now, except I dont go to school ofc'.

In *Technopoly*, Postman warns of the coming of a 'totalitarian technocracy'. He seems to have meant an actual political takeover, in which tech elites control the rest of the population with powerful tools that secure their dominance. Thirty-two years after his book appeared, this dystopia has not materialised, despite the continuous exponential growth in computing power that has led to vastly more potent machines. Today's tech moguls wield immense influence, yet are far more likely to face antitrust investigations than they are to seize dictatorial power.

While *Technopoly* focuses on society and systems, the deeper threat from technology today may be its effect on the individual soul. As virtual reality improves and AI chatbots grow more credible as friends and romantic partners, our tools are getting better at hacking our deepest human needs for love, purpose and adventure. If we let them, they will offer us ever-more-convincing simulacra of what we desire, while holding us back – through distraction, habit and lack of self-mastery – from experiencing the real thing.

Leonardo Ulian, *Techno Atlas 017(W) – Speak to Me,* electronic components and copper wire on mdf and paper, 2022.

risky flips at the lake thanks, in part, to the widespread availability of affordable smartphones, the very thing that drives the harmful trends Haidt highlights. The same large language model technologies that fuel worries about AI hold out the promise of even better treatments for future people with diabetes. We can't get the good without running the risk of the bad.

How, then, can we live well with tech? For starters, by doubling down on the analogue version of the very thing that tech products so often promise: community. Both Postman and Haidt, in different ways, help us to see that the challenge of tech is collective. Our response, too, must be collective; lonely acts of defiance, while often necessary, only get you so far. If, as seems likely, we're facing another round of Postman's 'ecological' change, we must become all the more active in nurturing flesh-and-blood communities that are robust enough to keep technology in its place. That can be as simple as a network of parents setting common norms for their families. Or it can be a circle of friends, a school, a company, a church or a commonwealth. Whatever the form, a strong communal culture can push back against pressure from technologies to shape humans in antihuman ways.

That's what my own community, the Bruderhof, is trying to do on a small scale. We embrace forms of technology that we believe help foster the flourishing of individuals and our common life, while acting together to set boundaries on forms that do not. To take the specific example of tech's impact on childhood, we practice a beefed-up interpretation of the guidelines that Haidt proposes in his book: Keep phones out of schools.

BUT THIS ISN'T AS INEVITABLE as Postman feared; we need not be at the mercy of our tools, nor despair of our capacity to control them. Clearly, it's necessary to be mindful of the ways that technology can shape us – of how (borrowing Marshall McLuhan's famous phrase) the medium can become the message. But those maxims capture only part of the truth. It is equally true that human beings are free to use our creativity for good. So long as we do so, our species' inventiveness is a feature, not a bug.

That's most obviously true in the case of medical technologies such as Tristan's CGM, that allows him to enjoy exactly the kind of 'play-based' childhood that Haidt urges as the antidote to a 'phone-based' one. A certain strain of conservative critique likes to draw the line between good and bad tech around medical uses, applauding progress in treating disease while seeing only the dangers of technical acceleration in other areas. Yet that tidy distinction breaks down in practice. After all, Tristan can practice

Leonardo Ulian, *Technological Mandala 59 – 1+1=3*, electronic components and copper wire on paper, 2015.

Don't give smartphones to kids until they're at least old enough to drive (Tristan is the obvious exception here, but his older brother is still phone-free and Tristan's phone is stripped down to essential functions). Keep kids off social media until they're old enough to vote, and in the case of adults, use it only when needed as a tool for work, study, or creative endeavours.

The same kind of boundary-setting is possible for society at large. In France, a panel commissioned by President Emmanuel Macron has recently proposed banning smartphones for kids under age thirteen and social media for kids under age fifteen. In regard to AI, even industry leaders agree on the need for more vigorous regulation than was typical in the early years of the internet and social media.

Communities with healthy cultures of solidarity matter then, at both micro and macro levels. Crucially, so do a community's goals in using the technology it has. The Book of Genesis, though written in an ancient context, sheds light on this. According to its first chapters, humankind is created 'in the image' of the Creator – endowed with creative ingenuity like that of God himself and commissioned to act as his representatives in the world. God sets the first human beings over the rest of nature as its stewards and caretakers, commanding them to 'fill the earth and subdue it, and have dominion over the fish of the sea and over the birds of the heavens and over every living thing that moves on the earth'.

After the first humans leave paradise, they set about doing just that, using their ingenuity to make tools, create structures and develop organisations. When they use these technologies to build rightly, acting in fulfilment of their divine vocation, their work can end up saving both humankind and other species from catastrophe; that's what Noah does in constructing the ark. But if they build in order to become independent of God, in violation of his commission to care for the world in his stead, their projects end in social breakdown. That's what happens to the builders of the Tower of Babel.

How can we live out our vocation as stewards of creation in the way we use technology? There are quite practical ways. Drawing again on what my community has learned, we need not just disciplines and restrictions, but also (even more importantly) positive guides for living. Crowd out the virtual with the real; be present in the physical world. Give the bulk of your attention to the people who are near you in person. Spend time outdoors; watch sunsets and moonrises. Plant

---

## A strong communal culture can push back against pressure from technologies to shape humans in anti-human ways.

---

vegetables, go birdwatching or fishing or hunting. Raise puppies and rabbits and pigs.

Thamus was right; with each technological revolution, a world ends. But a world also begins. Theuth's invention of writing brought the loss of oral traditions, while also making possible the composition of the Bible, not to mention the works of Plato, Dante and Shakespeare. Seventy millennia ago, a world ended and another began with the invention of the bow and arrow; no doubt the same will happen again when the next technological frontier is broken.

That's nothing to be afraid of, so long as we remember to proudly assert the freedom that belongs to humankind by right. We must remain the masters of our tools. For that, we need strong communities of the sort that many people around the world are striving to foster (some feature in this magazine). And we need to shape and use our tools in service of our human vocation, so that we can build, not Babels, but arks.

... עמו ... שרי בלק ... עם ...<br>
... ויקם בלעם בבקר ויאמר<br>
בלק לבו ... ויקומו שרי מואב ויבאו אל ...<br>
עמכם ויקומו שרי בלעם ... כי מאן יהוה לתת ...<br>
ויאמרו מואן בלעם מואבו ויבאו אל ...<br>
בלק שלח שלים רבים ונכברים מ...<br>
אל בלעם ויאמרו לו כה אמר בלק ...<br>
אל נא תמנע מהלך אלי כי כבד א...<br>
מאד וכל אשר תאמר אלי אעשה ...<br>
קבה לי את העם הזה ויען בלעם ויא...<br>
עבדי בלק אם יתן לי בלק מלא ביתו ...<br>
לא אוכל לעבר את פי יהוה אלהי ...<br>
קטנה או גדולה ועתה שבו נא בזה ...<br>
הלילה ואדעה מה יסף יהוה דבר ...<br>
אלהים אל בלעם לילה ויאמר לו ...<br>
לך באו האנשים קם לך אתם וא...<br>
הדבר אשר אדבר אליך אתו תע...<br>
בלעם בבקר ויחבש את אתנו וי...<br>
שרי מואב ויחר אף אלהים כי הו...<br>
ויתיצב מלאך יהוה בדרך לשטן ...<br>
לכב על אתנו ושני נעריו עמו ותל...<br>
את מלאך יהוה נצב בדרך וחרבו ...<br>
בידו ותט האתון מן הדרך ותלך ב...<br>
בלעם את האתון להטתה הדרך וי...<br>
יהוה במשעול הכרמים גדר מזה ...<br>
ותרא האתון את מלאך יהוה ותלחץ ...<br>
ותלחץ את רגל בלעם אל הקיר וי...<br>
ויוסף מלאך יהוה עבור ויעמד במ...<br>
אשר אין דרך לנטות ימין ושמ...

# From
# Scrolls *to*
# Scrolling

*At synagogue, the way we
read scripture has evolved.
Or has it?*

**J L WALL**

## I. THE SHUL

Walk into a synagogue on a weekday morning and you'll see three cultures of reading coexisting. Weekday services include a brief Torah reading, so at the front or centre of the room you'll find the oldest: the scroll of the *sefer Torah*.

Now look around: this is also the land of the codex. Congregants sit and follow the text of the Torah from the final pages of a siddur. Or perhaps they've pulled a volume from a shelf, and their eyes move from text to commentary to text. If I'm there, I've probably found something else. In the local Chabad House, there's a short selection of Rav Joseph Soloveitchik's essays that I'm slowly working through. Not distraction, exactly; just study of a different sort.

Then there are the phones. Hebrew letters flare, black against a blue-hued white, as someone nearby pulls up the prayers. He davens quickly and periodically pauses to wait on the rest of us before proceeding, a minute here, ten seconds there. During these waiting times, the screen switches over and his thumb scrolls, scrolls, scrolls through a social media feed. I can't see clearly enough to read what he pauses at. When I look around, I see he's not the only one.

## II. THE SCROLL

The earliest papyrus scrolls date back more than three millennia. In ancient Rome, public officials would read announcements from the vertically orientated *rotulus*. The one before us in shul is a *volumen*: that is, it rolls horizontally, right to left. It is pure animal product: roughly fifty sheets of hide parchment, sewn together with the hair or sinew of a kosher animal. Each sheet contains about five columns of forty-two lines each. The margins are exact: three inches at the top, four at the bottom and two at the right and left.

Once, all books, whether scrolls or codices, were handcrafted objects. Now, a *sefer Torah* must be, while the bound texts in the pews are mass produced. A single scribe or *sofer*, has laboured for a year, using a feather stylus to ink exactly 304,805 letters. He begins by lightly scoring horizontal lines across the parchment to guide his handwriting, then blesses his act and begins to write: slowly, deliberately.

Its reading is as ritualised as its production. The *ba'al kriyah* chants the words, vowelless and unpunctuated, according to the grammatically structured musical notations of Torah *trop*. Beside him, helping to hold the *sefer Torah* open, is the congregant who has said the blessing over the reading. At either end of the table on which the Torah lies, two more congregants stand, checking his words against the text in bound volumes. The reading takes at least three, often four people at a time – up to seventeen participants over the course of a Shabbat service. This is not private reading, but a group ceremony.

This scroll is the technology of the classical world, of Roman bureaucrats and Homer's steady temporal progression, once he was written down. Even its physical form implies continuity: it's difficult, a *process*, to jump from place to place. Judaism's rituals emphasise that continuity: though time progresses steadily from beginning to end, the narrative always circles back on itself.

On Simchat Torah, when the last lines of Deuteronomy are read, the congregation immediately returns to the beginning. The narrative has ended but the story has not, recurring in every generation: our father, the wandering Aramean; Pharaoh's hardened heart; we too were slaves in the land of Egypt. Political philosopher Michael Walzer has compared this experience of time to a spiral: history progresses, but also recurs. It's a temporal experience that, as Cynthia Ozick observes, allows

---

*J L Wall is a PhD candidate in English at the University of Michigan and is the author of* Situating Poetry: Covenant and Genre in American Modernism.

*Previous spread:* Ancient handwritten Torah scrolls from Yemen

for the metaphors that shape Judaism's moral imagination. 'As thyself', she notes, is the Bible's 'commanding metaphor' – a moral charge rejecting seeing the other as the Roman *hostis* and Greek *xenos* in favour of regarding them as the strangers and slaves we once were, or are. Such metaphor, she observes, relies on memory – the historical memory of the Torah. It is a moral imagination grounded in the rituals that allow the *sefer Torah* to contain past and future within the present.

## III. THE CODEX

The codex offers a different relationship to time. You can hold a book's pages, flip back and forth with ease. You can pause, put in a bookmark and come back later. You can read the ending first and know who the killer was all along.

The text of the Torah, writes Emmanuel Levinas, 'contains more than it contains … perhaps an inexhaustible surplus of meaning… Exegesis would come to free, in these signs, a bewitched significance that smoulders beneath the characters or coils up in all this literature of letters.' The codex, at least in Judaism, has become the technology of

this exegesis, transforming the ability to move back and forth in time into something else – a kind of simultaneity in which past and future once again find themselves contained in the present.

This was not always the case. The books we call the Talmud began as an oral tradition in which the human self, breath and bone, was the means of record and of exegesis. An oral Torah, Judaism maintains, was given to Moses at Sinai along with and as an elucidation of, the written Torah of the scroll. This was transmitted, studied and explained from memory, students and teachers swaying to the melodies of the law, until the destruction of the Second Temple. Its aftermath saw the dispersal of the Jewish people, the destruction of houses of study and the execution of great scholars. Preservation and continuity demanded writing.

Thus developed the Mishna in the form we know it today. The second-century labour of Yehuda HaNasi initially served as both a codification and a study aid, organising its precepts into six orders, sixty-five tractates and 525 chapters. All, of course, in manuscript:

The Sassoon Codex is one of the oldest biblical manuscripts, an 1,100-year-old leather-bound, handwritten parchment containing almost the full Hebrew Bible, the Tanakh.

handwriting. The Gemara developed during the next several centuries, records of the academies of Babylon and Jerusalem, their exegesis of the Mishna. Over time, the other voices that we find today on the pages of the Talmud began to speak, offering exegesis upon exegesis, though not, originally, on the same physical page.

---

**The effects of scrolling don't tell us the whole story, that of the history of reading and of the interactions between secular technology and sacred texts.**

---

Just as the *sefer Torah* transforms the scroll's steady progression to circularity, the Talmud insists not just that past and present *can* speak to each other, but that they always are. The Torah contains more than it contains – and *all* this, even the exegesis, was given at Sinai. Those conversations include voices ranging from the reign of the Second Triumvirate through late antiquity, arranged to argue, discuss and sling puns across time and space as if each knew or could anticipate the others. The codex or at least writing, is the technology that enabled the past to speak with the present, for the rabbis of the Diaspora to break bread with their dead.

This process was gradual. Manuscript codices facilitated exegesis alongside codification and early biblical manuscripts often contain brief textual notes on grammar, rare words, textual variants. But this was not their primary function.

The great change was the printing press. Gutenberg had printed his Bible by 1455. The Talmud as we know it began its life in 1520 when Daniel Bomberg started work on a complete printed Talmud.

So it came to pass that in the Venetian workshop of a Christian printer, the Talmud became the apogee of the codex, the most significant example of its *technological* difference from the scroll (and, indeed, of the printed codex from the manuscript). Today, we still follow Bomberg's imagining of what the printed page could be and do. The Mishna and Gemara appear at its centre, a column of continuous conversation enveloped in a ring of commentary: Rashi in his place of distinction, at the upper left-hand portion of each page, in 'Rashi script' – a font the great commentator never used, but which Bomberg chose to help visually distinguish portions of the page. Opposite him, the comments of the medieval French Tosafists, many Rashi's own students. Jewish printers gradually corrected Bomberg's textual errors and omissions, revising and adding commentaries to the pages. If you open a volume today, the page you look at will likely be a variation on the late nineteenth-century Vilna editions. Typesetters and the printing press reshaped the role of the codex in Judaism. Voices separated by thousands of years and miles greet each other on the page as if they have always been and always will be in conversation.

The Torah contains more than it contains – and that includes, you find when learning Gemara, your own encounter.

### IV. SCROLLING

The first transformation of 'scroll' from noun to verb occurred in the early 1600s. *To scroll* was to write in a physical scroll. Our sense of 'scrolling' doesn't appear until the 1970s, when we began scrolling through text on a screen in the smooth, frictionless and pageless manner. The metaphor is revealing in the way it misunderstands the scroll. There are, in a *sefer Torah*, if not pages, at least sheets. Their seams show. Our metaphor also forgets the physical act of reading. Even the Roman bureaucrat with his *rotulus*, scrolling vertically, as we do on screens, would feel his arms begin to tire and his grip loosen with the sweat on his palms beneath the sun.

Now we doomscroll through a text without beginning or end, with only the barest physicality: the touch of a fingertip. Scroll, scrolling: and ten minutes, half an hour have disappeared. We scroll to fill and then to kill, time. It's the nightmare inversion of Wordsworth's reveries, technology manipulating the mind's ability, in wonder or delight, to step outside itself to something perhaps higher. Our self-mesmerism lets us step outside ourselves – to nowhere.

If the scroll and the codex enable circularity and simultaneity, respectively, then scrolling emphasises ephemerality. Its conversations are brief and fleeting as breath – as the ragged and shallow *hevel* that the King James Version translates as 'vanity' in Ecclesiastes. Even critics of scrolling (I'm among them) begin to take on this quality. Our horizons are as brief as our lives – data trends across decades, perhaps extrapolating to gaze on a full century. But history, beginning with the first imprint of stylus on clay, is longer than we can fathom.

It's not that the effects of scrolling's ephemerality on individuals aren't important. They are. But they don't tell us the whole story, that of the history of reading and, especially, of the interactions between secular technology and sacred texts.

Step back into shul on a weekday morning: something is developing here, so slowly as to test the limits of perception. There's the scroll, the codices, and the surreptitious scrolling. It's a scene that suggests that even thinking about technology in decades-long cycles of creation, reaction and (ultimately) synthesis is deceptively shortsighted. More than a millennium passed between Yehuda HaNasi's codification of the Mishna in the early third century and Daniel Bomberg's 1523 Talmud. Three centuries more until the Vilna Talmud. That is the arc of time over which the People of the Book emerged from the People of Memorisation. So the question shouldn't be whether this era's innovation in how we read – scrolling through an endless 'text' produced by algorithms that, as seems likely,

read us more than we read them – is rejected or redeemed, but in what ways it comes to be synthesised with all that has come before it.

## V. THE SHUL (AGAIN)

Judaism is a religion of synthesis. It is also a religion of separation. The most famous passages of Ecclesiastes engage in this. I separate meat and dairy, wool and linen and, in ceremonies that bookend Shabbat, sacred time from the mundane.

So, step back into shul one more time, this time on a Shabbat morning. Here is the scroll; there are the codices. But there is no scrolling. 'The Sabbath is a palace built in time,' Abraham Joshua Heschel writes, a line so quoted it's nearly a cliché in Jewish communities. The walls built to protect this palace have, since the Industrial Revolution, made it sometimes seem like a fortress of Luddism. We must rest from creating – and modernity's innovations let us manipulate and recreate our world so effortlessly that we often don't notice it. This is the reason traditional Judaism prohibits not just things like cooking and writing on Shabbat, but also the active use of electricity. No phones, among other things. So, this palace is also one secured against scrolling.

Synthesis does not mean acquiescence. Judaism's synthesis with one technology, the codex, made it more fully itself. Its synthesis with many of modernity's technologies, the screens that enable scrolling no doubt now among them, has produced a different kind of transformation. Shabbat, more than a day on which we rest from creating, now serves as a bulwark raised to protect us – to keep us separate – from our technological creations.

If I'm to be an honest critic of scrolling, then I have to concede that the technology is *here* – it can't be unimagined. There is, I've read, a time for everything under the sun. But a time is not every moment. So as humanity navigates the synthesis of new ways of reading, we can also find ways to preserve space – and time. For those, we know, grant access to the sacred – and the human. ⤳

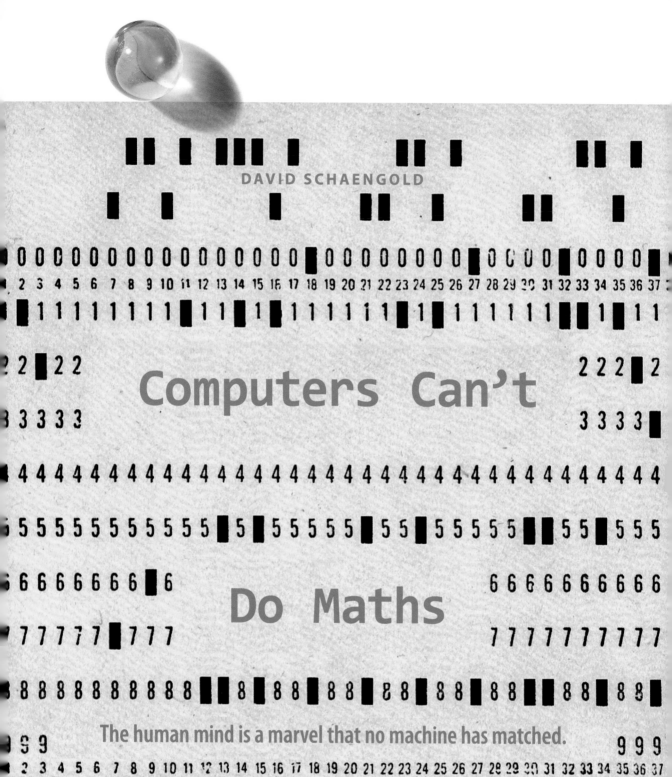

DAVID SCHAENGOLD

# Computers Can't

# Do Maths

The human mind is a marvel that no machine has matched.

An IBM punched card from the mid-20th century. Punched cards were pieces of card stock that stored digital data using punched holes.

0 0 0 0 0 0 0 0 0 0 0 0 0 0 0 0 0 0 0 0 0 0 0 0 0 0 0 0 0 0 0 0 0 0 0 ▮ ▮ 0 ▮
42 43 44 45 46 47 48 49 50 51 52 53 54 55 56 57 58 59 60 61 62 63 64 65 66 67 68 69 70 71 72 73 74 75 76 77

1 1 1                                                                                     1 1 1 1 1 1

2 2 2                                                                                     2 2 2 2 2 2

'SOME SAY THE WORLD will end in fire, some say in ice.' To these two venerable mechanisms of destruction, a third possibility has recently been added: apocalypse by artificial intelligence.

3 3 3    The idea that machines could become self-aware and turn against us has     3 3 3 3 3 3
been around for a long time, but it was in 2007, when I began to read a blog called *Overcoming Bias*, that I first learned there were people who took this possibility seriously.

4 4 4    Eliezer Yudkowsky, who has become the godfather of computer apoc-    4 4 ▮ 4 4 4
alypse terror, was a contributor to *Overcoming Bias* at that time. In his posts, he argued that it is existentially important for the human species to discover a way to 'align' future AIs with human values. If we fail to align

5 5 5    AIs, according to Yudkowsky, they will, by default, have some other goals,    5 5 5 5 5 5
incompatible with human values, or possibly human life.

6 6 6    Aligning a computer's values with our own is not a problem so long as    6 6 6 6 6 6
computers are dumb. Even an inkjet printer has 'goals', in some sense and

7 7 7    those goals may not be aligned with human values (printers in particular    7 ▮ 7 ▮ 7 7
may be actually malevolent), but this fact doesn't matter because the printer

8 8 8                                                                                     8 8 8 8 8 8

9 9 9 9 9 9 9 9 9 9 9 9 9 9 9 9 9 9 9 9 9 9 9 9 9 9 9 9 9 9 9 9 9 9 9 9 9
1 42 43 44 45 46 47 48 49 50 51 52 53 54 55 56 57 58 59 60 61 62 63 64 65 66 67 68 69 70 71 72 73 74 75 76 77

is deeply unintelligent. If computer programmes ever become more intelligent than humans, however, the divergence between their goals and ours will likely be fatal for the human species. Perhaps, to take an example from Yudkowsky's writings, they will have been trained by a heedless paperclip manufacturer to care only about creating paperclips more cheaply. This goal, pursued by a godlike intelligence that doesn't care about the poorly specified goals of its designers, only its own paperclip-maximising goals, implies the death of all humans, as a byproduct of the conversion of our planet into one giant paper-clip-production factory.

*Overcoming Bias* gave growth to an ecosystem of blogs and forums, and a small online community spent the 2010s discussing ways to align future super-intelligent AIs or, failing that, ways to defend ourselves against them. This community and its concerns remained niche until the beginning of the 2020s. Around that time, one particular approach to AI began to show astonishing results: machine learning, which relies on statistical patterns rather than deterministic inference. In November 2022, GPT-3.5 was released, a machine-learning AI model created by the company OpenAI. It was able to produce such humanlike responses to questions that suddenly the idea of a superhumanly intelligent computer programme began to seem intuitively possible to many more people. GPT-4, released to the public the following March, was even more apparently humanlike, and AI alignment worries went main-stream in a matter of months.

The general question – could a sufficiently capable AI pose a threat to human existence? – is not as easy to dismiss as one would hope. Experts in AI, including ones who work on the construc-tion of new models, are divided about how worried we should be, and some of them suggest we should be very worried indeed.

I do not propose to settle this question myself, but as I have read the various arguments on both sides, I have been startled to realise how close they sail to ideas I had not considered in detail since I was an undergraduate philosophy major. I realised, too, that some of those ideas might be relevant to whether we could successfully fight back against a superhumanly capable AI.

Whether a machine could be more intelligent than its human designers, I am unsure of; the term 'intelligence' is not used consistently in these discussions, in any case. But for many cognitive tasks ('task' feels less slippery than 'intelligence') I am inclined to believe that a computer could, in principle, be better than a human. There are some things, however, that I am confident a computer could never do. Maths, for instance.

THAT COMPUTERS CANNOT do maths is not very widely discussed. It is talked about in certain philosophy departments, and, naturally, it is considered with professional interest by computer scientists, though out of deference to their subject the conclusion is not usually put so bluntly. It is a constant source of frustration to computer engineers. It has not, however, reached popular consciousness.

The inability of computers to do maths is not merely theoretical. It poses a major practical problem in getting computers to do what people want them to do. A handout called 'Dangers of Computer Arithmetic,' from a computer science course at the Rochester Institute of Technology, for instance, notes several operations that are likely to cause problems, among them 'adding quantities of very different magnitude' and 'calcu-lating the difference of two very large values.'

---

*David Schaengold is a cofounder of* Radiopaper.com. *He is a philosopher and an architect, and lives in New York with his wife, Audrey, and their children.*

Great effort has been expended in hiding these realities from ordinary users. The impression given to casual users is that computer maths just works. But the underlying reality of 'just works' is a quite complicated substructure invented by clever humans, and reality sometimes slips through the cracks. Try typing '999,999,999,999,999 minus 999,999,999,999,998' into Google, for an illustration of how hazardous it is for a computer to calculate the difference of two very large values.

It's important to understand that these limitations are not bugs, in the ordinary sense of the word. Software deviates from its expected behaviour for many reasons, and bugs are only one kind of deviation: they are *mistakes*. A bug happens when a designer fails to consider some condition the software might encounter or forgets some important feature of the computer language the software is written in. The difficulty of performing certain mathematical operations (even ones that would be easy for a human) is not a bug, but an intrinsic limitation of digital computation with finite memory. What is missing in these cases is not due consideration, but *invention*. No way of executing certain maths problems computationally, with finite memory, has been *created* yet.

A certain tradition within the English-speaking philosophy world takes this point further, and claims – correctly, in my view – that computers can only ever *simulate* calculation, at best. They cannot successfully compute *any* function, even the simplest ones. They cannot, in other words, do maths.

The central argument in this tradition was made by the philosopher Saul A Kripke in his 1982 book *Wittgenstein on Rules and Private Language*. To explain his argument, I'll offer an example. Imagine you are a child, in first grade, and you have a best friend, Saul, also in first grade. You have seen Saul get a good grade on a quiz that tests single-digit addition skills, and you have seen Saul count up the players necessary to make a baseball team on the playground. In short, you believe you have observed Saul do addition in various contexts. But then, one day, you decide to pool your money and buy two gumballs, each costing 40 cents. 'We'll need 80 cents,' you say. '40 plus 40 is 80.' Saul gives you a puzzled look. 'No, we'll need 67 cents. 40 plus 40 is 67,' he says. 'What?!' you say. 'That's totally wrong. Think about 50 plus 50. That's 100, right? So, 40 plus 40 must be...' Saul shoots back, 'No, I don't know what you mean. 50 plus 50 is also 67.'

At this point you realise that Saul simply *does not know what addition is*. He got good grades on his single-digit addition test somehow, but it wasn't by doing addition. He was *never* doing addition. He was doing something else, and that something else, whatever it was, was not addition.

Kripke points out that machines are all like Saul. They can produce outputs that make it seem like they are doing addition, within a certain range, but in fact, they are only doing addition in the sense that we agree to treat them as if they were doing addition. They are *simulating* addition. Even when they get the answer right, they are not doing maths, just as Saul was never doing addition.

Computers simulate maths very usefully, of course. Imagine Saul is a strange kind of savant and can do any addition-like problem instantly, so long as the result is less than a million. If the result is more than a million (the result according to addition, that is), he says the answer is 67. You still wouldn't say that Saul understands addition or is even *doing* addition, but he'd certainly be useful to have around when you're doing your maths homework.

The simulation is so skillful, in the case of computers, that we forget there is an extra step and round it off as if the computer were doing the real thing. But the computer cannot act as we do. It does not know the difference between simulating and doing, because it cannot do anything but simulate. As the philosopher James F Ross wrote, following Kripke:

> There is no doubt, then, as to what the machine is doing. It adds, calculates, recalls, etc, by simulation. What it does gets the name of what we do, because

it reliably gets the results we do (perhaps even more reliably than we do) when we add... The machine adds the way puppets walk. The names are analogous. The machine attains enough reliability, stability, and economy of output to achieve realism without reality. A flight simulator has enough realism for flight training; you are really trained, but you were not really flying.[1]

Computers depend on their designers, in other words. They do not, themselves, do math, though their users can do maths *with* them.

There is one sense in which computers can 'do maths'. They 'do maths' in the same way that books remember things. A book really does have a kind of memory. Stendhal's *Memoirs of an Egotist* contains some of his memories. But it does not contain them in the same way that Stendhal's own mind contained them. These analogous ways of speaking are harmless in everyday life, and probably unavoidable, but faced with genuine uncertainty about the dangers of AI, we should learn to make finer distinctions in our speech. If the librarians of the New York Public Library were regularly issuing warnings that the Main Branch might turn homicidal and rampage through the city sometime in the next few years, I would want to be quite careful in ensuring that no false analogies had crept into my thinking about what books are capable of. Faced with a potentially dangerous AI, we should carefully examine our ways of speaking about computers too. To say that computers can do maths is a reasonable analogy, but speaking as unmetaphorically as possible, it is not true.

1. James F Ross, 'Immaterial Aspects of Thought', *Journal of Philosophy* 89, no 3 (1992): 136–150.

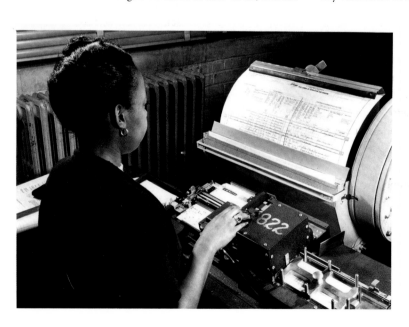

A clerk creates punched cards containing data from the 1950 United States census.

**A** NATURAL RETORT to all of the above is that, if computers can't do maths, then neither can we. After all, humans also have limited memory, and frequently produce wrong answers to maths problems. A natural retort, but not a sustainable one. Admittedly, it is not *impossible* that our minds work the same way Kripke describes machines working. The view contains no internal inconsistencies. In the same way, it is possible that the physical world is an illusion, that other people do not exist, etc. But even more than these views, the belief that humans do maths the same way computers do leads to absurd conclusions, as I will explain.

My kids sometimes ask me how high I can count. I've noticed that they stop asking this question once they reach a certain age, usually around six or seven. This is because the question does not make sense once you understand what a number is. If there's a single highest number you can count to, you don't really grok numbers. The difference between computers and humans doing maths is a bit like the difference between the younger kids who think that 'how high you can count' is a real thing and the older kids who have successfully understood how numbers work.

This seems to be hard for some people to accept. In discussions with friends around this question, the principal difficulty seems to be understanding the nature of the work that a human is doing when interacting with a computer. The *representation* of the numbers that occurs only in the mind of the human is conflated with the *execution* of a particular programme that takes place in the computer.

We anthropomorphise habitually. We do it to all kinds of things, imputing emotions to our smoke alarms and intentions to stuffed animals. To computers, we impute a kind of reasoning power that they cannot have. We're able to do it because we ourselves are able to pivot so effortlessly from abstraction to reality. We say things like, 'Oh, given infinite memory,' etc., and instantly we're in the world of purely abstract objects, of stuff that lives, as far as we can tell, only in our minds. To us the transition between the theoretical and the actual happens almost instantly and unnoticeably. It happens so quickly and we do it so often that we don't realise it's magic. But it really is magic, in the sense that it's amazing and we have no idea how it works, and computers could never ever do it.

Think back to the subtraction problem I mentioned earlier. 999,999,999,999,999 minus 999,999,999,999,998. How do you know the answer is 1? Why is it obvious? If you're like me, you visually scanned the digits and noticed that they were all the same except for the last one. Given my understanding of subtraction, it's clearly safe to ignore all those extra digits and the problem reduces to 9 minus 8.

How did I know that this is a valid way of doing subtraction? I don't think anyone ever taught me this method. Even if they did, I haven't just memorised this method as one of several procedures for performing subtraction. I can just see that it is correct and that it will give the same result, if used correctly, as any number of other procedures I might use.

You could, of course, programme a computer to use this same method (and in fact, Wolfram-Alpha, one of the most sophisticated online calculators, is able to do something like this). The method itself is not special; what is special is being able to recognise the validity of the method. I recognise its validity because I have learned the *concept* of subtraction, which transcends any particular method of calculating subtraction.

Despite thousands of years of philosophising about the human mind, we do not have a detailed mechanism-level understanding of how it is that a human might come to have something like a concept or even exactly what one is. Our current inability to understand what a concept is, however, does not mean that the difference between what a human mind does and what a computer does is mystical or vague. The difference itself is quite clear.

I'll try to explain the difference more concretely. Imagine you have some marbles in a bag. You take 2 of them and put them on your desk. You count 2 marbles on your desk. Then you take 2 more marbles out of the bag and put them on the desk. Now you count 4 marbles on the desk. Have the marbles or the desk done any maths? Does the movement of the 2 extra marbles from the bag to the desk cause there to be a 4 in the world where there was previously only a 2, or is the difference only in your head?

(This is a rhetorical question; the difference is only in your head.)

Let's keep going. Now you want to know what 4 plus 4 is. You start taking marbles out of the bag – but *oh no!* There were only 3 marbles left in the bag. You have taken all 3 out and put them on the table, but you know you need to put 1 more thing on the table to get all the way to 4 plus 4. Fortunately, you have a pencil in your shirt pocket. You take it out and put it on the table with the marbles and make a mental note to yourself to count the pencil as well as the marbles. You count 7 marbles, but you remember your mental note and count the pencil too, and so you manage to get to 8. Phew!

The math that's going on here is not in the marbles or in the pencil. It is some faculty of the human mind that binds the marbles and the pencils together into 8 things.

Computers can be programmed to treat pencils as well as marbles as number-counters, but they cannot be programmed to represent *anything arbitrarily* as a counter. Computers have no target beyond the counters they actually have. What they can count *simply is* what they can represent.

If that were the way it worked for us, there could be missing integers. There could be an integer between 5 and 6, for instance.

The idea seems absurd. The numbers 5 and 6 are *defined* as the fifth and sixth integers. It's a

contradiction in terms to think that there could be another integer between them. This is, of course, correct. There is no integer between 5 and 6. But if you're a computer, you cannot be sure of this fact.

One common way for computers to store numbers is in a format called 32-bit float. Like all numeric formats used by computers, 32-bit floats can only represent certain numbers. The number 16,777,217, for instance, cannot be represented as a 32-bit float. (It is the smallest positive integer that cannot be represented in the format.) The previous integer, 16,777,216, and the one afterwards, 16,777,218, can both be represented, but not 16,777,217.

If you imagine a computer that simply stores all numbers as 32-bit floats, 16,777,217 just *does not exist* for that computer. No calculation that requires storing that number will work quite right. If you ask such a computer to take 16,777,216 and add 1, it will not be able to give you the result. Depending on the details of the algorithm, it will probably either return the original number or skip ahead two numbers to 16,777,218.

In practice, computers do not simply store all numbers as 32-bit floats. And various algorithms make it hard (though not impossible) to find simple integer patterns like this that your laptop cannot handle. Nonetheless, no matter how many layers of abstraction you add, and no matter how sophisticated you make your algorithms, every digital computer has integers it does not and cannot know about. These integers are, for the computer, inexorably and absolutely missing.

And if these integers are truly missing, they might be anywhere on the number line, as far as the computer knows. They could be in the single digits. If your mind did maths the way a computer does, there could be an integer between 5 and 6 and you would never know it. No matter how hard you tried, you could never count using this number or perform operations that required you to use this number, even in an intermediate state. You would be doomed to skip this number when you counted. To you, going from 5 directly to 6 would seem right, just as going directly from 16,777,216 to 16,777,218 would seem right if you were a computer that only used 32-bit floats.

In this situation, maths would still seem perfectly complete because you would always get some answer or another to every problem you thought about. But your answers would consistently be invalid. If correct, they would be correct only by coincidence. In other words, unless there is some profound difference between the way humans do maths and the way computers do maths, maths is basically fake. That's a hard pill to swallow. It is much easier to believe and indeed much more likely to be true, that computers can't do maths, and humans – even though we don't know how – can. Computers and humans both have finite memories. But we humans somehow do something, in spite of that limitation, that takes the infinite into account when we do maths. Computers do not.

C AN THIS SAVE US from the AI monster? This is the speculative part.
　　A computer programme that takes over the world needs to be able to act in the world. To act in the world, it must have an internal representation of the various situations or states, the world can be in. These states must each reduce to some individual number, however large or however stored, that corresponds to the memory state of the computer for that representation. If (so the concern goes) a hypothetical AI is so much better than humans at understanding the world in this 'full internal representation' sense, what hope do we have to fight it?

So far so good, as far as the AI is concerned. The people who study Turing machines (an abstracted, formal model of computers used in computer science) might tell us that the whole universe is 'computable' in the sense that you could choose a system of representation with distinct numbers for every state the universe can be in. Further, you could perform operations on these numbers. The so-called Church-Turing-Deutsch principle suggests, speculatively, that in a quantised universe (one in which energy and matter are not continuous but are broken up by reality itself into discrete chunks, or quanta), any physical process whatsoever has at least one precise mapping to a computable function.

Computable, yes, but not by any actual computer. 'Computability' is an abstraction, and computers are not abstract entities. What if a world state maps to a number that a computer cannot represent? Suppose this state maps to 16,777,217 and the computer only stores 32-bit floats? The computer, no matter how sophisticated otherwise, is completely blind to that state of the world. It cannot imagine or reason about that state.

What does this look like in practice? It looks like SolidGoldMagikarp. This word, if you can call it that, describes a creature from the Pokémon franchise and it proved to be indigestible by GPT-3.5. If you typed it into ChatGPT, the chatbot

interface OpenAI offers for some of its models, it would react in unpredictable and odd ways. Often, it would simply treat the word as if it were in fact 'distribute.' I typed the following phrase into ChatGPT recently, in fact: 'Tell me about SolidGoldMagikarp.'

In response, the chatbot replied: "'Distribute' can refer to several different concepts, so it would be helpful to know what context you are asking in. Could you please provide a bit more information or clarify what you are looking for?'

This is not an isolated example. Users excited by SolidGoldMagikarp quickly found a number of other strings that also resulted in odd, non-sequitur outputs from GPT-3.5.

SolidGoldMagikarp was fixed in GPT-4, and also in GPT-3.5, as far as I can tell. And whatever weird logic caused it probably lived at a much higher level of abstraction than 32-bit floating point numbers, in any case. But this sort of thing is exactly what it looks like for a computer to be blind to certain world states and no number of abstraction layers can prevent such situations from arising again.

This is a concrete prediction: for any machine intelligence instantiated on a digital computer, there will always be SolidGoldMagikarps. There is no way, in principle, of eliminating all such conceptual blind spots.

The trick is finding the blind spots. I don't have any process to recommend. But we can be sure there are world states beyond the comprehension of any AI. And I suspect those world states will not necessarily be ones that seem extreme to us. We won't have to reverse the orbit of the moon. It will be a matter of odd, seemingly incomprehensible phrases. Or donning cardboard boxes, as some US Marines did recently as a training exercise in trying to evade an AI-backed camera setup. The boxes might as well have been cloaks of invisibility, as far as the AI was concerned, and the Marines strolled right past the camera. It's as if the boxes shifted the world state representation to a hidden integer, and in so doing the Marines simply vanished from the conceptual apparatus of the computer.

We are used to human intelligence, but whatever capabilities a computer might have, intelligence is not one of them. Even a machine that could out-negotiate, out-strategise, and generally outwit us can still be undone by certain oddly specific inputs. The human mind is magic, or might as well be, and it is by this magic that we can defeat the AI. Not by outwitting it or by unplugging it, or any such thing, but by the sheer weirdness of the human mind. ➤

# Blackberry Hush in Memory Lane

*after Sylvia Plath's "Blackberrying"*

*Nobody in the lane, and nothing, nothing but* poison ivy
along its shoulders, hunched into fields, once stripped, clear
back when the Black Angus herd intervened;
sold off to cover a single semester of college tuition,
their rasps of papillate tongues became meat, sliced
in rasping bawls, no longer licking

all those leaves of three, let them be
in the lane, witness the itch, down deep green
the gloss of encroachment on hallowed ground
in fricatives vining, chafing up every tree,
every gasping gap of old fenceline, edging the lane
to a climax, choking

the blackberry canes that used to be
in the lane, where we lapped
all the summers' juices, lavished
for us to grow and sing, young glistening things
facing the fall and cull, another year older, we'd run
that quarter mile stretch to catch the school bus, slick

potholes to dodge in the lane, the grey mist before day,
while the cows would wake and aggregate, round
flanks and eyes shining, dark as blackberries,
to curtail the poison ivy, close-cropped;
later, quietly ruminate the cud—
so tender the release

JENNIFER FAIR STEWART

Artwork: Louis Prang & Co., *Poison Ivy from the Plants
series*, lithograph, 1862–69 (detail).

# Prison Parenting

*Even a little tech can make a big difference.*

ROBERT LEE WILLIAMS

THE LIGHTS ARE OFF in the cell block, but I'm lotused on my bed, my tablet in hand, angling the screen to catch the lamp's orange glow. I'm scrolling through the 60 new pictures that my 15-year-old daughter, Harmony, just sent me. All of the pictures that now dominate my photo gallery are snapshots that span her life from toddlerhood to adolescence. Looking at everything I missed as a father is bittersweet. Prison took the best years of my life and kept me from being present during those tender, irreplaceable moments.

I'm a resident of Sullivan Correctional Facility, a maximum-security prison in Fallsburg, New York, in the Catskill Mountains. The first decade after I was incarcerated, this technology wasn't available to me.

Ironically, at that point, I was about to raise her full-time when I went away. Harmony's mother, Sarah, has had problems with addiction for many years. In 2009, my probation officer, Dan Bryant, said the Department of Social Services contacted him and told him Sarah had lost custody of Harmony and that I had to get custody of her.

I simply said, 'OK'. I didn't want her to fall into the clutches of a broken foster care system. Unfortunately, while in the process of getting custody, I committed the crime that would separate us for a long time, a crime I eventually had to tell my daughter about.

In June that same year, I killed my girlfriend. We partied that night, drank, danced, then argued on the drive home. Once we were inside the apartment, she stabbed me. I stabbed her. She died. I went to the hospital, then prison. I was sentenced to 25 years for manslaughter.

Many people are kicking heroin, crack, or meth addictions when they first get to prison. I was a social media junkie kicking the World Wide Web cold turkey. Unlike the old-timers who'd been in prison since the 80s and 90s, when pagers and Motorola flip-phones were considered the

height of technology, I knew why I was craving it. In 2010, when Facebook and Twitter were in their infancy and Myspace dominated the web, I had a recurring dream of logging in to Myspace, pointing and clicking and typing messages to family, friends and exes – my inbox bursting with replies.

Then I'd hear metal tapping metal, a guard's baton banging against my bars. I'd wake up locked away from lightning access to everyone who mattered. Back to perfecting my penmanship on love letters and deciding between purchasing a box of Banquet chicken on commissary or purchasing ten stamps to write home to loved ones. I'd feel guilty when I chose drumsticks.

When you first enter prison, there is a two-year institutional adjustment period. That is the average amount of time research shows a prisoner

*Robert Lee Williams is an incarcerated writer in New York. His work has been published by the Prison Journalism Project.*

needs to adapt to institutional rules and restrictions, and learn the ropes of prison's subculture.

After four straight hours of being bounced around a janky-smelling bus – cuffed and shackled to strangers – dumped into a dystopian world of walls and bars, you put your head down and develop a steely gaze if you want to survive. In the beginning and sometimes in the middle of a long prison stint, prisoners do what we call 'just bidding'. It's the time period when a sort of inertia sets in (because prison is monotonous and depressing): you lay back, smack squiggly lines from your 13-inch TV that's always on the fritz, cook ramen, pump iron. You drown your days and nights in prison must-reads, like Alexandre Dumas's *The Count of Monte Cristo*, Sister Souljah's *The Coldest Winter Ever*, and self-help books like Rhonda Byrne's *The Secret*, or Eckhart Tolle's *A New Earth*.

But it's also critical for your mental and emotional wellness that you maintain communication with your friends and family, your support system, a connection to your life before. Otherwise, you tend to forget who you are and you become what cages create. It helps to find a loyal friend who will reach out to estranged family and friends who aren't visiting or reaching out to you.

My friend Krissa was the first of many people kind enough to take on that role. After I told Krissa I hadn't heard anything from Sarah, she went on Sarah's Facebook profile, printed out the pictures of Harmony that Sarah had posted on her page and mailed them to me. In one picture, which was taken in the Galleria Mall, Harmony is in a pink dress, her hair scrunchied into two fuzzy brown puffs, wearing a smile that makes me wonder if she's smarter than the photographer. I taped it to the wall – against the insane-asylum white, she was the reference point for my sanity. It was all I knew of her for years.

N 2015, I WAS ON a rare phone call with my mom. Rare, because the phone system was so expensive: I couldn't call her often. She asked me if I had heard anything about my son, Ameer, or Harmony. I hadn't heard anything about my son and the things I heard through the grapevine about Harmony didn't sit well with me. I hated when my mom brought up the topic of my children. It knotted my chest.

I didn't answer. I just sighed a full breath into the receiver.

I was sick of the phone system costing too much for my poor loved ones and the fact that hardly anyone I knew took the time to write letters by hand. Besides that, I didn't want to deal with the anxiety I felt whenever I thought about involving the legal system in what I believed was a private family matter; especially when I knew that if I had internet access at my fingertips, I wouldn't have these problems.

'Robert,' my mom said, 'don't lose track of your children while you're in there.'

'You're right.'

In truth, I wanted to cross the emotionally taxing bridge of parenting upon my release in 2030. Prison is stressful enough. Was I ready to

meet the added challenge? As an adopted child who never knew his biological father, I understood that my selfishness was rooted in fear and would be a disservice to my daughter. My mother's statement gave me goosebumps that lingered for days.

Shortly after we talked, I filed the documentation with the courts to apply for visitation with my daughter. In the meantime, as I waited, I quelled my anxiety with magical thinking.

I imagined swaggering into the visiting room and introducing myself to Harmony as her father. Her light-brown face would light up; so would mine. I'd pretend she was a rocket ship as I lifted her high above my head. I'd make blast-off sound effects and tell her, 'Don't worry, Daddy's here.'

About a month later, a correctional officer called me to my gate to sign for legal mail. I scribbled my name. The officer tore open a letter from Family Court, stuck an eyeball inside it, then handed it to me. The court couldn't serve Sarah the papers. Her mailing address had changed.

Sarah's Facebook profile hadn't changed, however, so I had my longtime friend Keith go to her page. He told me that it had been inactive for almost a year. He sent her a message anyway.

I had no idea where my daughter was, who she was living with, whether or not she was safe, or how I would find her. I wondered if she hated me. I would have understood. But as a man, that gnawed at my masculinity.

Inside prison, when guys run out of war stories to tell each other to establish their place in the social hierarchy, they don't have much on the outside to brag about except their women and children, so those relationships become status symbols. I watched men around me receive letters from their kids. I received zilch. I wished guys, 'Have a good one,' as they strutted out to visits and returned showing off Polaroids. In Greenhaven Correctional Facility, the room where we held our prisoner-run NAACP [Black civil rights] meetings acted as a safe space for my Brooklyn buddies to boast about their

children graduating from high school or college. Out of respect, I appeared the best of listeners.

Secretly, I brimmed with envy. In this context, my envy had a positive effect – it sparked a growing fire within me. I wanted to be a father to my baby girl and for her to see me as more than just another incarcerated Black man.

By this time, I'd read PEN America's *Handbook for Writers in Prison* (which initially made me believe becoming a writer in prison was possible), devoured Arthur Plotnik's *Spunk & Bite*, and made Andrea A Lunsford and Robert Connors's *The St Martin's Handbook* and Joseph M Williams's *Style* my new North Star. When I located Harmony, I didn't want to just be proud of her. I wanted to show her something she could be proud of me for, too. And I knew it somehow would involve slinging ink.

By 2019, I'd been living in Clinton Correctional Facility, a maximum-security prison in Dannemora, New York, for three years. On the outside, Clinton was famous for the escape of David Sweat in 2015. Inside the New York prison system, Clinton is notorious for its barbaric violence.

I watched prisoners get their faces slashed by

In July 2019, the New York State Department of Corrections introduced the tablet program, which was intended to give the incarcerated population access to music and movies, e-books and a secured emailing platform to communicate with friends and family. The latter, in particular, has been shown to lower the recidivism rate.

In Clinton, the kiosks were installed inside the slop sinks (which look as disgusting as they sound). The slop sink is a closet-sized cell containing an industrial sink, used to wash out mops and rags and hang brooms and dustpans on crooked hooks and nails. The slop sink is creepy, dark and infested with roaches. And I'm a big-time roach-a-phobe.

Luckily, you don't have to type your emails directly from the kiosks. The email app on your tablet allows you to type up your message and save it in a draft folder for later editing, or put it in your outbox. When you sync your tablet to the kiosk, it sends any messages you have in your outbox and any incoming messages appear in your inbox.

I didn't waste any time. I got in touch with Harmony's aunt Shayna, who brought me up to speed in a series of emails: 'I feel so sorry for that little girl.'

Harmony had been adopted. Sarah was no longer allowed supervised visitation with Harmony, but Harmony communicated with both of them and her little cousins by phone. I begged Shayna to tell Sarah how to set up a JPay account, the email platform used in prison.

But I would have to wait a full week for a reply. According to the rules, each prisoner is assigned to sync his tablet to the kiosk once a week, for 15 minutes. Initially, when it came to opening our cells to use the kiosks, the correctional officers gave us a hard time. Their collective sentiment: they didn't want us to have tablets.

I wasn't going for that. The instant satisfaction of sending an email and knowing someone else receives it right away, is powerful. Plus, I needed

other prisoners, with tiny bits of hand-sharpened metals and plastics – in the hallways, in the mess hall, in the cell blocks. The best and safest place to use the phone was in the gym, which you only had access to once every other day. Otherwise, you were only permitted to use the phone in the recreational yard.

Clinton's yard is set up like the Roman Colosseum, with a flat area in the centre. It's full of sand. The sand pit is surrounded by a large hill, crisscrossed with dirt walkways and stone steps, where small areas called courts are sectioned off with plywood fences. You want to stand on a court, not on the flats. Nine times out of ten, whether it is a thousand degrees in the summer or below zero in the winter, stand on those flats and you're liable to get caught in the middle of a gang riot, forced to choke on pepper spray spewed from gas canisters that guards shoot from towers to disperse the crowds.

The hostile landscape was unavoidable. If I wanted to work out under sunshine or snowfall, or make a phone call to decompress, I had to go outside. I found myself spending more time in the cellblock, banging away on my Swintec typewriter, writing fiction and poetry, wishing for a safer way to communicate with the world beyond the razor wire.

to send out emails to pull my family together. I decided to risk a disciplinary infraction that could potentially get my tablet taken away by doing what we call 'stealing syncs'. This means syncing your tablet to the kiosk without a correctional officer's permission.

I'd tuck my tablet into the waistband of my pants and walk slightly hunched all the way to the mess hall, hoping it wouldn't slip. I'd seat myself at the table that the officers point at to go back first. On the way, I'd make it my business to stand in the front of the line. Once we got back inside the cell block, I'd race upstairs ahead of everyone else – including the officer whose job it was to open our cells. I'd slip into the slop sink with the kiosk, plug the USB into my tablet, wait for it to sync. 'Come on, come on, come on,' I whispered in the dark, looking around for roaches, listening for footsteps coming up the stairs. 'Come on, come on, come on!'

Then I'd sneak out of the slop sink and blend in with the rest of the prisoners walking to their cells to lock in.

O N DECEMBER 15, 2019, Sarah messaged me, gushing: 'Harmony looks just like you! Lol… your skin tone, your eyes… she's like a girl you!'

I read Sarah's message a thousand times and just knew that the joy I felt inside my cell that day would sustain me until she put me in contact with Harmony.

But that day took another several months to come. On 21 July 2020, Harmony, now 12, logged into Sarah's JPay account: 'Hey… it's Harmony. I can't wait until I can see you. I'm sorry I don't have a lot to say, but I love you and always will. I don't know if it will show, but I sent a picture of me. OMG. I didn't see the send button at the top.'

That night, I stayed up until three in the morning tapping away: 'Harmony, I detest sounding cheesy and Darth Vader but – this is your father and I love you.'

I decided that we wouldn't start our relationship based on lies, because once children know you've lied to them, they grow up resenting you. I told her why Sarah and I broke up and that I didn't abandon her. I told her exactly what I was in prison for, my reflections, why it happened, and why I should've never done what I did. I told her how I used to be well known in the neighbourhood as a rapper named 5 Starrz, but now I was a writer and planned on self-publishing a book. I thanked her for saying she loved me, because I feared she might hate me. I called her my little Beauty Queen and signed off: 'I hope this letter lifts your spirits! You are no longer alone. I love you with every light speck of my soul.'

Harmony responded: 'When I saw that you thought I hated you, it made me so sad. I never hated you, or my mom. When I was younger I did wonder why things were how they were but I never once ever said or thought of including the word hate next to anyone I love.'

Since July 2020, we have become constant pen pals. I've learned to decipher the terse acronymical language in her messages. She told me her favourite rapper was Pop Smoke, created hairstyles to complement outfits, and demanded I take her seriously 'all the time'. I looked forward to her long stream-of-consciousness messages about everything and everyone that ticked her off, or depressed her, which she titled, 'Just Venting'.

Her ventings, I confess, sucked me into some dark places where I felt the kind of powerlessness that gives you purpose. Reading of the many injustices my daughter suffered at the reckless hands of what were supposed to be responsible adults, from such a young age, I felt so much guilt and rage. I could've taken it out on a fellow prisoner who didn't deserve it, or a correctional officer, which would have done nothing to help. Instead, I channelled my anger. I shared with her the story of my own challenging childhood and the despair of prison, and how I understood so much of what she was going through.

I wrote to her, 'I understand what it's like to feel trapped, alone, with no one to talk to but the page.'

AT ONE POINT, I FOUND MYSELF running out of stamps. When we first got the tablets, a single stamp cost 46 cents. By comparison, in New York, the wage for prison labour is 10 to 65 cents an hour, so stamps cost us a lot. Since then, the cost of a JPay stamp in New York has been lowered to 12 cents per stamp, thanks to families and advocacy groups complaining of the high prices. Advocates continue to argue that the prices for secured messaging services like JPay should be lower, or even free for prisoners, given the social benefits.

I appreciate these concerns and certainly acknowledge that in some states the price for JPay's services are higher still. But this is not at the front of my mind when I think about JPay. Rather, I'm grateful to the Department of Corrections for bringing in this company and tablets to provide us with services. Sometimes I wish the tablets weren't so cheaply made and glitchy, and that lengthy messages didn't take days to receive. (This adds a layer of difficulty when I write for publications and try to work the text out with my editors!) Maybe it would be better if it weren't a monopoly, and another company was brought in to make

the prices more competitive. This is corporate America; in order to function and provide a service, someone has to make money. More to the point, whatever the provider or the cost, I would not have been able to form a bond with my daughter without this technology.

I still worry about my son, Ameer, now 24, whom I have tried to reach via similar methods without success. I'm still hoping to connect with him one day. I also want him to know I love him.

Meanwhile, I began to fulfill my dream of becoming a professional writer. In 2023, I landed three pieces: 'In Prison, Networks of Addiction Run Deep,' published by the Prison Journalism Project, 'Turning Sentences Around,' published by PEN America and 'Good Writing in a Bad Place,' published in *Literary Hub*.

I wrote all three of those pieces on my tablet and then worked with editors on them via emails. This tablet enabled me to build a career and forge professional connections that I can cultivate upon my release. Now, when Harmony googles her dad, even though I'm locked up, she sees more than just a prisoner. She sees a writer with long dreads. Over the phone, she tells me I still have 'drip' (style). She is proud enough of what her old dad has made of himself to show her friends my YouTube presentation about higher education in prison, which was recorded live via Webex. (It felt good to tell her that I was a Black historymaker: to my knowledge, the first incarcerated New Yorker to do this.)

Scrolling through another set of pictures Harmony sent me, where she looks all grown up, I can't help but take pride in thinking that she does look like a girl me. But it's even better for Harmony, in my opinion, that she also looks like her mother. Most of all, she now knows that I love her and have always loved her, even during those years we weren't able to reach each other.

For me, JPay helped reunite a family. You can't put a price on harmony. ➤

# Give Me a Place

*An East Tennessee farmer praises a simple piece of technology – the rock bar.*

**BRIAN MILLER**

I CAN'T SAY WITH ANY CERTAINTY that wielding a rock bar has contributed to my moral improvement or made me a better citizen – I'll leave that for others to determine (hopefully out of my earshot) – though I will say that a quarter-century of living this farming life has brought so many benefits, among them an old way of seeing.

Christmas gifts on our 50-acre East Tennessee hill farm follow a predictable pattern. One year

it's a bundle of work gloves, the next, a maul for splitting wood. This past Christmas it was a lightweight metal rod with which to stabilise the PTO-driven auger when drilling postholes. It replaced the old rod, twisted serpentine from heavy use – curious but next to useless. The presents I receive typically offer up an augury of future farm activities.

I still recall my first Christmas on the farm in 1999. That was the year I received the rock bar as a gift from my partner, Cindy. The rock bar is the tool that Archimedes may well have had in mind when he postulated, 'Give me a place to stand, and I shall move the earth with it'. It's a 20-pound, six-foot-long iron bar with a round flat head on one end and a wedge on the other, the perfect combination of form and function. If you are looking for a quick means to separate the men from the boys, put a rock bar in a fellow's hands, then step back and observe. Your average muscle-bound gym rat will last about 30 minutes with the rock bar (and indeed most farm work); that wiry farm boy can use it all day. Turns out the farm kid is the 'real' man of the two.

Having dug hundreds of postholes, I can confirm with some authority that the rock bar lives up to its name. When you have dug through two feet of clay only to hit solid, seemingly immutable rock, the rock bar is the only tool to part the sea of stone. Raise it high in the air, wedge side poised and bring it down with full force. Repeatedly. Breaking big rocks into smaller rocks with the bar gives you a feel for what it must be like for the prisoner in a Russian gulag: it's hard, make that *very* hard, physical work, yet when it's used voluntarily it's intensely satisfying.

Once your hole is dug and your post is in place, flip the bar over to the round edge. As dirt is added to fill the hole, use a rhythmic pounding action with the rock bar to compact the earth. Tamping requires short brutal strokes around all sides of the post. No substitute tool or action is as effective in seating an eight-foot wooden anchor.

When it's not pulverising rocks into shards, the rock bar moonlights as a lever. Is there a stock trailer that needs to be shifted, a boulder that needs rolling uphill or downhill? In the right hands, this tool will do the job and with minimal effort on your part. Seldom does a day go by without my bringing out the rock bar.

I learned about the rock bar long before I embarked on farming. My first exposure to its wonders came in the early 1970s, working alongside my brother and cousin on my uncle's cattle ranch, in the Big Thicket across the Sabine River in Texas. His family lived in Beaumont, a gritty industrial city on the Gulf Coast an hour from my Louisiana home. Most weekends my uncle and cousin could be found either hunting squirrels with their Catahoulas or tending their cattle. Once, when pressed by my cousin's mother why my uncle and cousin were always heading to the woods or the ranch, my uncle told my aunt, 'Cille, you can't raise a boy on a hundred-foot-square city lot'. For a few summers in a row my uncle 'educated' his boy, along with me and my brother, in the art of digging holes, setting posts and firming them with the rock bar in the East Texas dirt. Lift and pound, lift and pound. So, in 1999, when I unwrapped an unusually heavy and long Christmas package, I knew and embraced, what lay in store.

A good tool is more than just the sum of its parts (or, in the case of the simple iron rock bar, the lack thereof) and it is only with repeated application that the tool's merits are fully revealed. If a tool is chosen well, the physical act of using it often has the unexpected benefit of bonding you to a place. Much like the carpenter who can point with pride to the house he built, having sweated and set hundreds of wood posts with a rock bar, then stretched and stapled a fence to each and every post, you find that you have built a connection with your land, your place – a firmed-up appreciation, you might say, one that is tangible to the eye in the results stitched across a pasture.

The work done on this farm of mine has shaped me physically, strengthened me, sometimes hurt me (as my aching shoulders remind me most mornings). I expected those particular

---

*Brian Miller has farmed in East Tennessee since 1999. He and his partner, Cindy, raise sheep and pigs. He is the author of* Kayaking with Lambs: Notes from an East Tennessee Farmer.

consequences as I learned to work the land. Beyond the physical, I also had an expectation, even a hunger, that this farming life would reshape me in other ways. Although I couldn't have predicted the specifics, I certainly carried the hope that it would lead me away from the rootless and easy path I was travelling and towards an older way filled with purpose, competency, values and even virtue.

Farm work provides daily if not hourly opportunities for deep introspection, chances to glimpse links within links, the cycles unending, like a skipped rock that sends ripples across the water. A hole is dug, a post is set, the earth is tamped, a fence is stretched and gates are hung. A flock of sheep is released to graze and lambs are born, weaned, reared and butchered. On a late summer's day, the ram is turned in among the ewes and the cycle repeats anew. Grass is sown and hay is cut. A nephew visits, helping work the farm each summer, learning the hows and whys of the rock bar. You pass on the anecdote to him, then later to his mother (your sister), that you can't raise a boy on a hundred-foot-square city lot. The cycle repeats. He develops muscles, not only the physical kind but also the mental ones that he did not know he possessed. The rock is skipped again and the waves begin another journey across time and waters.

It is not just this tool – a straight iron bar wielded by a pair of hands doing simple work – that has the power to shape a farm, reshape an individual and help make this place productive. It is that by working modestly and manually you can find the right lever to shift the circumstances of your life. Then maybe on some land of your own, in a workshop or in your garden, things begin to move. If you are lucky, one day those changes will lead to where you should have been and who you were meant to be – where the blinkered blindness of this modern life falls from your eyes and you can see your way back into the world. ➤

*Gerard Manley Hopkins* (1844–89) is recognised as one of the greatest English-language poets, though none of his poems was published in his lifetime.

**Julian Peters** is the author of Poems to See By: A Comic Artist Interprets Great Poetry *(Plough, 2020). He lives in Montreal, Canada.*

AND FOR ALL THIS, NATURE IS NEVER SPENT;

THERE LIVES THE DEAREST FRESHNESS DEEP DOWN THINGS;

AND THOUGH THE LAST LIGHTS OFF THE BLACK WEST WENT

OH MORNING, AT THE BROWN BRINK EASTWARD, SPRINGS—

# TV in Allentown Becomes Reality with WFMZ-TV

## Channel 67 Will Feature Many Live, Local Shows

# Machine Apocalypse

## Will AI bring the end of the world or is it already here?

**PETER BERKMAN**

ELON MUSK THINKS THE APOCALYPSE is on the way. He's not the only one. In the eyes of Silicon Valley's leading experts on technology, an existential threat to civilisation approaches: artificial intelligence. In their best-case scenario, AI heralds unprecedented mass unemployment. The worst: our machines get smart enough they don't want us around any more. Academics at some of the world's top universities share Silicon Valley's uncharacteristic discomfort with technological change: ethicists, philosophers and historians have all weighed in at industry discussions on the impending catastrophe. Given the end-times rhetoric, it's curious that religious thinkers have been left out of the conversation. Two neglected examples – Marshall McLuhan and Romano Guardini – might have especially pertinent contributions to make. Both men would have agreed with Silicon Valley that the scale and speed of technological progress could only appropriately be described as apocalyptic. But McLuhan and Guardini, despite never meeting in life, would have added the same qualification: the end of the world is already here.

McLuhan, a Canadian academic, died in 1980 at age 69, just as his trendsetting study of the information society, *The Medium Is the Message*, was bringing him to the peak of his reluctant celebrity. He never met Guardini, an Italian-born German Catholic priest and theologian who died in 1968 at age 83. Guardini's ideas have quietly set the tone of Catholic thought for decades; every pope since the 1962–65 Second Vatican Council has cited his work as a major influence. McLuhan's own influence, by way of contrast, has been confined largely to secular audiences, including Musk himself.

McLuhan's deep Christian faith pervades his writings, even if it barely enters into his reputation as one of the 20th century's major media theorists. Yet both McLuhan and Guardini, despite their very different backgrounds, looked at the new technological order unveiled in the aftermath of World War II and came to the same conclusion: the world – at least as they knew it – was ending.

In fact, it already had. Guardini's *The End of the Modern World* (1956) warned that we stood on the threshold of a new era in human history. 'Something has come up that has not existed before,' Guardini told an audience of students at the time, 'the unity of inhumanity and machine.' Something essential had taken place: the authority of the state converging with the power of technology. Human arts, once a means of security from natural dangers, had become the source of danger itself. In the optimistic atmosphere of postwar reconstruction, Guardini was a rare voice of alarm. This seemed a striking contrast to his prewar liturgical and spiritual writings. In the eyes of his critics – and many of his admirers – Guardini went from being ahead of the times to falling behind them. Even Guardini's biographers tend to relegate this 'melancholic' period to something like an emotional spasm, a footnote to his theological masterworks. There's an element of truth in this. Guardini, who spent the interwar years mentoring young people, was persecuted by the Nazis in the 1930s and watched the horrors of World War II and the Holocaust unfold from a kind of internal exile. Yet as much as these events made Guardini disillusioned with modernity, his 'melancholic' critiques of technological man nevertheless drew on his earliest work, just as McLuhan's did.

In the 1925 book some consider his masterpiece, *Der Gegensatz* (opposition), Guardini resurrects the ancient and medieval theory, shared by

*Peter Berkman is a scholar of the work of Marshall McLuhan and Romano Guardini and the frontman of the chiptune band Anamanaguchi. He is a research fellow at the Center for the Study of Digital Life (CSDL) and an editor of CSDL's* Dianoetikon: A Practical Journal.

Aristotle and Saint Thomas Aquinas, that there are two kinds of knowledge: a 'sensible' knowledge akin to a sort of maternal instinct and a 'formal' knowledge that formulates everything for exact repetition. Like Aristotle and Aquinas before him, Guardini argues that 'there is nothing in the intellect that was not first in the senses'. Intellect requires intuition; power requires ethics. One kind of knowledge presupposes the other. And the confusion of these two different forms of knowledge, he says, is the reason we so often uncover hidden biases in apparently 'neutral' scientific thought. 'The individual stands in a topsoil.' The more hidden the bias, the more aggressively it asserts itself.

In *The End of the Modern World*, Guardini looks back across the age of the Enlightenment and sees the gap between the scientific and the human widening to a chasm; what remains of civilisation teeters on the edge of the abyss. In the order he saw emerging from the catastrophe of the war, Guardini discerned the contours of our contemporary dependence on technology. We have become physically all-powerful, Guardini warns, but our moral capacities have withered. Our reach outstrips our grasp.

Our capacity to control the world around us has itself become uncontrollable, reducing both 'divine sovereignty' and 'human dignity' to words in a history book. 'Contemporary man has not been trained to use power well,' Guardini writes, 'nor has he – even in the loosest sense – an awareness of the problem itself.' Blind to the crisis our staggering technological capabilities represent, we conceive of everything in terms of power. But the remedy is more of the disease. 'Again and again one is haunted by the fear,' Guardini concludes, 'that to solve the flood of problems that threaten to engulf humanity... only violence will be used.' We have begun to revert to idol-worship, Guardini writes, 'the worshipping of tools'. The 'new culture' Guardini saw emerging in 1952 would be defined, above anything else, by a 'single fact': danger.

In the optimistic days of the immediate postwar era, Guardini's darker vision of the future promised by scientific rationalism was far from common. It was one shared, however, by Marshall McLuhan. 'I'm not an optimist or a pessimist,' McLuhan declared in an interview; 'I'm an apocalyptic.' His first book, *The Mechanical Bride* (1951), was the culmination of a six-year project to understand the psychological weaponry of American advertising – and disarm it. McLuhan had clipped out hundreds of advertisements from

---

We have become physically all-powerful, Guardini warns, but our moral capacities have withered. Our reach outstrips our grasp.

---

newspapers, comics, and magazines and set them beside short essays, what he saw as a practical (if not moral) task of attempting to render the subconscious products of the culture business intelligible. Although McLuhan's strategy was necessarily defensive, he was convinced opponents of new technologies of manipulation had to explore and understand their enemies, not simply retreat into denunciation. He suggested that the Americans trapped in the culture business adopt the attitude of the whirlpool-menaced sailor in Edgar Allen Poe's 'A Descent into the Maelström,' seeking 'amusement in speculating upon the relative velocities of their several descents towards the foam below'.

As McLuhan told his interviewer, he wasn't a pessimist about the world around him. But he wasn't optimistic either. The 'dominant pattern' of advertising, McLuhan wrote, was 'composed of sex and technology', eerily reminiscent of the future the 19th-century novelist Samuel Butler anticipated in his pioneering science fiction book *Erewhon*. 'Machines were coming to resemble organisms', McLuhan wrote of Butler's fiction

---

*Artwork in this piece:* Newspaper clippings and advertisements for 1950s TV and radio networks, along with pictures of Marshall McLuhan and Romano Guardini, superimposed on coloured AI-generated illustrations.

and our reality, and 'people… were taking on the rigidity and thoughtless behaviourism of the machine'. This mechanistic worldview, that McLuhan called 'know-how,' inculcated by humans' ever-increasing use of machines with 'powers so very much greater' than their own, was altering culture at every level. In the long term, he speculated, it threatened to make human life 'obsolete'. In the present, it manifested itself in the victory of force over thought in the most immediate way imaginable: 'murderous violence'.

Like Guardini, McLuhan came to a new understanding of the contemporary world through his study of medieval thought. In his PhD thesis, written during the war and sent to the University of Cambridge chapter by chapter to avoid destruction by U-boat, McLuhan reread human learning in the Western world through the lens of what medieval thinkers considered the foundation of a good education, the 'trivium' of grammar, dialectics, and rhetoric. The church baptised the classical ideal of the *doctus orator*, the learned speaker, interpreting

structures through grammar, convincing others through words (rhetoric) and using dialectic as both an art for settling disputes and a means of intellectual speculation. Studying modern advertisers, McLuhan realised something key to his later work: intellectual progress isn't unidirectional, but a pattern of repeats and retrievals, backwards and forwards across time. Advertisers in the 1950s were using the same methods of persuasion as 13th-century rhetoricians. But this technique was amputated from the rest of the arts: an instrument used without understanding, conveying countless effects without communicating ideas. And in a consumer society, this error determined the whole shape of our culture. The medium was the message.

Though working along different lines, McLuhan saw the problem of our age in similar terms to those of Guardini. Technology has given an untrained humanity unprecedented power over itself. At the same time, we have become numbed by this power, using it – or being used by it – without purpose or thought. Writing to

Catholic philosopher Jacques Maritain in 1969, McLuhan posed the resulting crisis in dramatic terms: 'There is a deep-seated repugnance in the human breast against understanding the processes in which we are involved. Such understanding involves far too much responsibility for our actions'. He was echoing the thesis of another Guardini book, *Power and Responsibility* (1961): we have the first, but lack the second. As a result, our creations overpower and overcome us. Narcissus didn't fall in love with himself, McLuhan warned, reinterpreting the ancient Greek fable. He fell in love, fatally, with an image of his own creation.

Similarly, for Guardini, the civilisation of technique, all means without ends, threatens to erase the 'essential values of nature and human work', destroying reverence for the past – and for the person. Guardini's 'topsoil' had blown away. Herein, he writes, 'lies the terribly new'. 'Moral injustice,' 'the spirit of violence,' 'cruelty' all existed in the past. But the technological society, by 'eliminating the personality of the human being,' dismissing all considerations other than those of raw power, 'achieves something that is even more terrible than evil': a system of thought and action with ethics left entirely out of the picture. In a strange way, both men understood the world in terms not entirely dissimilar to Musk and other contemporaries, who agonise over humankind being replaced by unfeeling robots or prophesy our enslavement to artificial, inhuman logics. Only for McLuhan and Guardini, this wasn't a possible future, but a real present. The robots aren't coming. To paraphrase Alasdair MacIntyre, they've been ruling us for some time.

What can we do? How do we escape the apocalypse? We can't escape, McLuhan and Guardini would argue, and it's pointless trying. There's no way back. But there is a way through. Shortly after the publication of *The Mechanical Bride*, McLuhan carried out a research project comparing the effectiveness of learning by television, in person, over radio and through the printed word. Assessing the results, television came out ahead and print at the bottom of the list. Advertising agencies circulated the results, McLuhan recalled, with the assertion that 'here, at last, was scientific proof of the superiority of television'. The satellite was mightier than the pen. The advertisers' reaction was 'unfortunate', McLuhan later wrote. They 'missed the main point'. The study hadn't indicated superiority, but difference, 'differences so great they could be of kind rather than of degree'.

'The new mass media', McLuhan writes, aren't instruments, but contexts, 'new languages, their grammars as yet unknown'. Each language in the 'ecology' of telecommunication 'codifies reality differently; each conceals a unique metaphysics'. None of these 'languages' are neutral; each perpetuates particular sensitivities and particular blindnesses. 'There is no harm in reminding ourselves that the "Prince of this World" is a great PR man, a great salesman of new hardware and software, a great electrical engineer', the master of an environment that is 'invincibly persuasive when ignored'. Left unwatched, the reigning technological paradigm will inevitably provide our customs and habits for us. When the environment is invisible, the path of least resistance is determined by the medium itself. Our plans, as Guardini puts it, plan us. In order to recover human agency over technique, the invisible environment has to be made visible. 'Apocalypse', McLuhan says, 'is our only hope'.

'THE FIRST ADAM simply looked at things and labelled them.' So McLuhan observed in 1968. Christ, the second Adam, was a maker. Sharing in Christ's inheritance means we have, McLuhan added, 'the mandatory role of being creative' and not passive. To survive the end of the modern world, we need to exhibit the capacity of artists to remake and reshape their surroundings – a faculty, McLuhan noted, that they share with saints. For Guardini, the antidote to an 'autonomous technical-economic-political system' is the virtue of humility, that the medievals considered to consist, like most virtues, in a kind of balance. On one hand, we must admit that we do not have total control despite what inventions we may devise, without abnegating our God-given powers as human beings, namely those of understanding and action. 'Jesus' whole existence is a translation of power into humility. Or to state it actively: into obedience to the will of the Father as it expresses itself in the situation of each moment.' It is through self-renunciation – the acts Christians once described as 'mortification', dying a death to self, remaining aware of our human limitations in an inhuman, Promethean order – that we can see that order transformed. The sensors and satellites are not our enemies. They are us, the products of human intelligence, operated by human will, extensions of our powers and our frailties. And where sin abounds, grace also abounds. Christ calls us to let the light of God into the world – including the parts of it we ourselves have made.

McLuhan's solution is consonant with Guardini's. As a form of humility, we must admit that technology's influences on our own perceptions are not immediately obvious. McLuhan saw his own research as a sustained effort to avoid self-deception: 'study the modes of the media in order to hoick all assumptions out of the subliminal, non-verbal realm for scrutiny and for prediction and control of human purposes'. The goal is understanding. 'What is the alternative to violence?' an audience member once asked McLuhan. He answered instantly: 'Dialogue'. We can't pierce the veil of the 'hallucinogenic world' we exist within if we rely on 'a reactionary romantic attitude,' rejecting science and technology to return to 'a state of nature that could never be realised.' We can only do so in action, in the conscious and considered rejection of possibilities we might otherwise passively accept. 'An ascetic is necessary, that is, a willingness

to renounce technical achievements, so that higher values are maintained.' 'It is not brains or intelligence that is needed', McLuhan writes, but 'a readiness to undervalue the world altogether'. Such a way of life, however, would appear as 'anti-social behaviour', he warns. Making the invisible assumptions of our age visible might make Christians appear as odd as refusal to sacrifice to the gods once was. But it could also, he speculates, begin a 'religious renaissance' in the world of the machine.

In 1959, speaking to the first-ever students to study computer science at Munich University, Guardini expressed a similar hope for the future. It was far too late to stop the change that was the end of the modern world, he thought, but it didn't have to be a change for the worse. Moral agency could reassert itself over the technological order, in the life of individuals and of humanity as a whole. Guardini suggested a 'spiritual council of the nations, in which the best from all political areas would consider these questions together'. He told the students that 'it is time for a new virtue' to meet the challenge. 'A living consciousness of humanity' could enable them to comprehend the 'whole of our existence, with a truly sovereign objectivity'. A restoration of the lost 'topsoil' of human life, a spiritual development on the scale of our scientific progress, an ethic to constrain and correct technique: this may sound like a utopia, Guardini granted. And in history, 'the creative and unifying forces work more slowly than the one-sided violent ones.' Nevertheless, he maintained, utopias have often enough become reality. And all things are possible through God.

Pessimism about technology's role in human life pervades our society as much as technology itself does: Silicon Valley's contemporary fear of an AI-initiated cataclysm is a stark contrast to the arcadia the Valley's founders imagined they'd create. But although McLuhan and Guardini counselled caution to the hopeful, they'd suggest realism in place of fear. Rather than rushing to

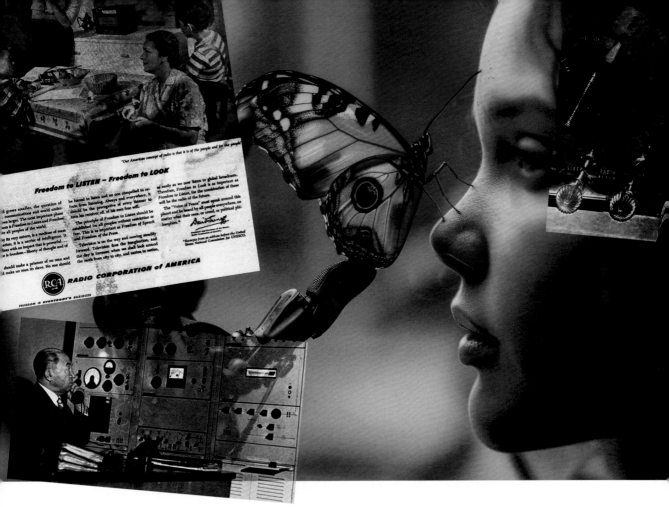

find the right answers, they'd say we should try to find the right questions. There's more to technology than what we perceive on the surface, good and bad. There's more to human beings too. And it's only when we understand the second that we can comprehend the first.

If McLuhan and Guardini were correct about technologies reshaping human sensibility, it seems fair to say that we are only at the beginning of the transition from an electric age to a digital age. It took centuries for medieval man to become modern man and the shift from electric humanity to digital humanity is perhaps the most rapid and sweeping change we have ever undergone. Our response to this, as individuals and as a church, must be not to search for enemies – robotic or otherwise – but to look instead to the ways, old and new, that God is making himself known to us through the world he has made. The combined memorative power of all the world's networked computers, equipped with hundreds of billions of sensors, with the interpretative power of code painstakingly labouring to make concrete all manner of human bias, as we live today, stands in literal comparison to six words uttered at a table in Jerusalem 2,000 years ago: 'Do this in memory of me'. Only one of these remains when the power goes out. Explaining why he called himself an apocalyptic, McLuhan noted that both optimism and pessimism are secular states of mind. 'Apocalypse is not gloom', he said. 'It's salvation.' ⤳

# Send Us Your Surplus

*South Sudan's kids thank you for that shipment
of hip-joint ball bearings.*

## MATTHEW LOFTUS

I**T WAS ALWAYS A BIG DEAL** when a new shipping container arrived at our hospital for women and children in South Sudan. Hospitals or charitable organisations would donate supplies from time to time; it took months for the containers to travel across the world, clear customs and then arrive by truck. We eagerly unlocked the big metal doors wondering what was inside: Gauze? Non-expired drugs? Surgical drapes to replace the ones that were fraying from constant use?

One day, our hospital staff opened up a container and found that a hospital in America had donated boxes and boxes of... ball bearings for artificial joints. These materials are crucial for joint replacement operations, but our hospital was not capable of performing such surgeries; we exclusively delivered babies and took care of very sick children. One of the boxes somehow got opened and some of the children from a nearby orphanage took to playing with thousands of dollars' worth of titanium in the red dust of their football pitch.

Hospitals throughout the developing world depend on donations from the West. Photos courtesy of Global Health Ministries.

Our team leaders called it 'the blemished lamb syndrome,' referencing the passage in Leviticus: 'You shall not offer anything that has a blemish, for it will not be acceptable for you.' Donations to charities and impoverished countries are often the leftovers, the tax write-offs and the unwanted flotsam of somebody's inventory. In African markets you find T-shirts celebrating the wrong Super Bowl winner or emblazoned with sponsors of the 2012 Turkey Trot from some small American town. In any mission hospital you will likely find a room full of junk collecting dust; someone brought it over thinking it might be useful and it's been sitting there ever since. Missionaries and locals alike laugh as they judge the people who have somehow given the opposite of the widow's mite.

Admittedly, it can be hard to know what might conceivably be useful to someone somewhere far away. Our family has seen clothes and furniture that we threw away pulled out of the rubbish pile and put to some kind of use. We do try to give away clothes that we've outgrown and things we're no longer using, but it feels shameful to offer a friend a shirt that has a hole in it. Apparently, it's a little less shameful to pull such a shirt out of the bin. But whether trash or treasure, a gift of firstfruits this is not.

More uncomfortably, the topic of unwanted donations brings to mind summer mission trips to Mexico where team after team of gringo teenagers put on Bible Club skits all summer long or large-scale food aid that props up American farmers while undercutting African ones. There is an entire literature describing 'helping without hurting' – it goes well beyond the scope of dumping one's junk on someone else and shows how even the best-intentioned gifts can go wrong. Fundamentally, though, the basic problem we

often face is that people want to feel like they have done something good without taking the time to ask whether that good thing is as blessed to receive as it is to give. And it is easiest to give away the stuff you weren't going to use anyway.

The excesses and rubbish of our world rightly make us uneasy: giant patches of plastic waste in the ocean, enough food to end world hunger tossed into skips, electronics we trade in after two years of use dumped on African beaches. As a doctor, I think of the American nights on call when I ordered thousands of dollars' worth of unnecessary tests as a rearguard against personal-injury lawyers. These are symbols of technology's long shadow: every single-use plastic syringe has to go *somewhere* when you're done with it. One of the awful revelations of the last few years has been learning that most plastic simply cannot be recycled and that plastics manufacturers deliberately try to obscure this fact.

'When you invent the ship, you also invent the shipwreck; when you invent the plane you also invent the plane crash', French theorist Paul Virilio famously said. 'Every technology carries its own negativity, which is invented at the same time as technical progress.' The absurdities of excess are yet another manifestation of this law. Excess is technology's gratuity, a sign that we've gone above and beyond what was necessary to survive. We can use that excess for things we don't necessarily need, since producing abundance allows us to think about more than sustenance. If leisure is the basis of culture, excess is in many ways the basis of generosity. Our ancestors spread their fields with night soil, fermented their leftovers and used every part of the animal because doing so helped them survive; wasting anything could decrease their family's chances of survival. Nowadays, we have so much excess that 'decomposing over

---

*Matthew Loftus lives with his family in Kenya, where he teaches and practises family medicine. He is also a regular contributor at* Mere Orthodoxy. *matthewandmaggie.org*

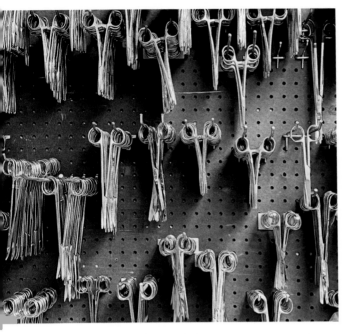

simply thrown away after it is no longer usable or fashionable. And we buy a lot of stuff that could easily be called 'junk', wasting the precious money that we are supposed to steward. Wastefulness, like most other vices, only grows stronger the more it is indulged.

The phenomenon of non-fungible tokens (NFTs), that I can still only think of as 'Beanie Babies for people who think they are smarter than everyone else', illustrates how excess can produce its own stultifying impetus to have more for the sake of having more. When you invent the internet, you invent the inane million-dollar picture of a monkey. When you invent the plastic bead, you invent the beanbag chair, the Beanie Baby, the microplastic pollution and on and on it goes. At least NFTs pollute the air with the carbon it took to make them and then disappear; some Beanie Babies might still be floating in the Pacific when Jesus comes back.

centuries in a landfill' is an inevitable part of the life cycle of almost anything we use. There's no escaping it, either: almost anywhere you go around the world, you will find even the most remote villages littered with plastic rubbish and discarded car parts.

In an ideal world, we would have the same 'use it up, wear it out, make it do, or do without' attitude of our ancestors, just with more comfortable margins. The excess that the incredible technologies of the present have created would be put to good use, whether that means giving it to those in need or carefully recycling it. (The givers would ideally ask recipients what they wanted before giving it away.) Dorothy Day's maxim about your extra coat belonging to the poor would be taken seriously, if not literally. Somehow, we would find a way to take all the money that gets wasted on unnecessary MRIs in America and give it to my patients in Africa so that 40-year-old mothers don't die of cervical cancer.

In reality, we not only throw away a lot of waste, we produce a lot of stuff knowing that it will be

A T THE SAME TIME, all of us in mission hospitals who cluck our tongues at the givers of blemished lambs do so without making too much of a racket. After all, we all depend on donations of money and equipment from people who have chosen to be generous for a wide variety of reasons. The same container that brought the ball bearings had lots of useful stuff in it and we were able to get a fair amount of treasure out of someone else's trash. We would not be able to do what we do if there were not an enormous amount of excess.

My career as a medical missionary wouldn't be possible without technology: the pharmaceuticals, the lab tests, the single-use plastic syringes and the other day-to-day materials are underpinned by other systems such as computerised insurance authorisations delivered over mobile phones or the aeroplanes that fly my family and me back and forth to Kenya. In the hospital, we take a much more aggressive approach to making do

Global Health Ministries collects second-hand surgical instruments for hospitals overseas.

and wearing out, reusing many 'single-use' items (when it is safe to do so), but we still have to throw a lot away at the end of each day. Yet my life would not be possible without the excesses technology has given us; if my friends and supporters were all subsistence farmers using every part of the animal, there would be nothing left to buy my plane tickets or our plastic syringes.

I personally cannot say that I always offer the Lord an unblemished lamb myself; my time of prayer is interrupted by a really funny thought I need to post on social media or my day of missionary service may be punctured by an angry outburst about a broken piece of equipment. (Last week a critical surgery was cancelled because the hospital had only one stent for a procedure that required two and I was ready to hurl the iPad I was holding out the window.)

Thrift is a virtue that can be cultivated, but it, too, can be deformed: Ebenezer Scrooge was not a wasteful man. His inability to let any margin be gleaned is the approach we don't want to take. Nor do we want to be like Judas, grumbling about the cost of perfume for the Lord's feet. We will always have the poor with us, which means that, as the Levitical law instructs, we must always find a way to leave behind the gleanings for the needy. One imagines, perhaps, that more than a few blemished lambs found their way to a poor man's table after being rejected at the temple.

OUR APPROACH TO THE PROBLEM of waste should be threefold: to cultivate a sense of gratitude, to count the cost and to give until it hurts.

Gratitude forces us to look at what we have as a gift from God. If you look at what you own as what you are owed, you are likely to decide that you deserve more. If you pretend that your disposable income is entirely under your control, you'll buy a lot of stuff you simply dispose of. Thanking God for everything – your food, your home, your clothes and even your toys and other frivolities – gives you freedom from possessiveness. And when what you have is a gift, you feel less inclined to hold on to that excess for yourself.

Counting the cost mostly means living with your waste. Every week the hospital where I work runs its incinerator and the acrid smell of burning plastic wafts over our living quarters. It's an ever-present reminder of what it takes, technologically speaking, to save lives, a negative externality that many Westerners get to avoid. I suspect that most people would throw less away if there was a law ensuring it was burned within a mile of where it was disposed of. Try burning your own rubbish for a week and see if it changes the way you consume.

The widow gave her last two coins, but we cannot expect society to run on that same level of moral fortitude. I like what Dorothy Day has to say a lot, but I put myself into a mental health crisis when I tried to run my life with a fraction of her scrupulosity. We are each called to sacrifice differently for the sake of God's kingdom, but sacrifice applies to everyone and is achievable no matter your station in life.

I think that when it comes to managing your resources, you should give until it hurts at least a little. God wants your heart more than he wants your wealth, but excess wealth is like too much plastic packaging wrapped around your heart. Give more of it away, expose yourself to some of the vulnerabilities that excess shields us from and offer to God slightly more than you are able to spare from your flock.

So, then, do not store up for yourself rubbish on earth, where the microplastics do not disintegrate and you depend on the bin lorry to take your stuff away. The abundance generated by technology will always try to tempt you if you can't restrain it. Rather, be grateful for the excess and put it to good use, for there are many who need it. But for heaven's sake, please check with us first before you ship it to Africa. ➤

# Masters or Slaves?

*Four writers reflect on the purpose and
power of technology.*

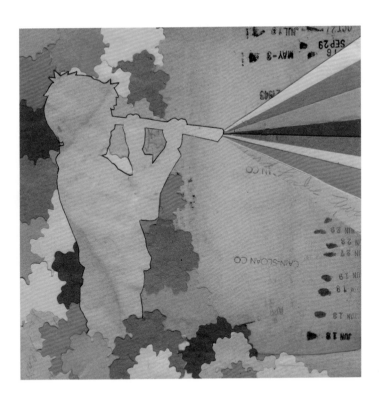

*E F Schumacher (1911–77) was a German-British statistician, economist and writer.*

THE DOMINANT MODERN BELIEF is that the soundest foundation of peace would be universal prosperity. One may look in vain for historical evidence that the rich have regularly been more peaceful than the poor, but then it can be argued that they have never felt secure against the poor; that their aggressiveness stemmed from fear; and that the situation would be quite different if everybody were rich...

This dominant modern belief has an almost irresistible attraction, as it suggests that the faster you get one desirable thing the more securely do you attain another. It is doubly attractive because it completely bypasses the whole question of ethics: there is no need for renunciation or sacrifice; on the contrary! We have science and technology to help us along the road to peace and plenty and all that is needed is that we should not behave stupidly, irrationally, cutting into our own flesh. The message to the poor and discontented is that they must not impatiently upset or kill the goose that will assuredly, in due course, lay golden eggs also for them. And the message to the rich is that they must be intelligent enough from time to time to help the poor, because this is the way by which they will become richer still.

Gandhi used to talk disparagingly of 'dreaming of systems so perfect that no one will need to be good'. But is it not precisely this dream which we can now implement in reality with our marvellous powers of science and technology? Why ask for virtues, which man may never acquire, when scientific rationality and technical competence are all that is needed?

E F Schumacher, *Small Is Beautiful: Economics as if People Mattered* (Harper & Row, 1973), 23–24.

Hollie Chastain, *Future*, 2010.
*Opposite:* Hollie Chastain, *Harvest*, 2017.

*Antoine de Saint-Exupéry (1900–44) was a French writer and aviator who was killed in action during World War II.*

THE MACHINE does not isolate man from the great problems of nature but plunges him more deeply into them.

Numerous, nevertheless, are the moralists who have attacked the machine as the source of all the ills we bear, who, creating a fictitious dichotomy, have denounced the mechanical civilisation as the enemy of the spiritual civilisation...

It is hard for me to understand the language of these pseudo-dreamers. What is it makes them think that the ploughshare torn from the bowels of the earth by perforating machines, forged, tempered and sharpened in the roar of modern industry, is nearer to man than any other tool of steel? By what sign do they recognise the inhumanity of the machine?

Have they ever really asked themselves this question? The central struggle of men has ever been to understand one another, to join together for the common weal. And it is this very thing that the machine helps them to do! It begins by annihilating time and space...

It seems to me that those who complain of man's progress confuse ends with means. True, that man who struggles in the unique hope of material gain will harvest nothing worthwhile. But how can anyone conceive that the machine is an end? It is a tool. As much a tool as is the plough. The microscope is a tool. What disservice do we do to the life of the spirit when we analyse the universe through a tool created by the science of optics or seek to bring together those who love one another and are parted in space?

Antoine de Saint-Exupéry, *Wind, Sand and Stars*, trans Lewis Galantière (Harcourt Brace Jovanovich, 1992), 43–45.

Hollie Chastain, *Gilbertville Public Library*, 2009.

*Jean-Pierre Dupuy (b 1941) is a French engineer, writer and philosopher.*

THE PRINCIPAL DANGER facing humanity, I believe, is the temptation of pride. The fatal conceit is believing that technology – which has severely impaired all those traditional (that is, religious) systems that serve to curb the tendency to excess, itself inevitably a part of human action – will be able to assume the role that these systems once played when the capacity to act bore only upon other human beings and not upon nature. To believe this is to remain the prisoner of a conception of technology that sees it as a rational activity subject to instrumental logic, to the calculus of means and ends. But today technology has much less to do with fabrication (poiesis) than with the power to act (praxis), which now means: the power to unleash irreversible processes; indeed, the power to generate 'out-of-controlness'. In abandoning ourselves to scientistic optimism, counting on technology to rescue us from the very impasses into which it has led us, we run the risk of producing monsters that will devour us.

Jean-Pierre Dupuy, *The Mark of the Sacred*, trans M B DeBevoise (Stanford University, 2013), 29–30.

*Hannah Arendt (1906–75) was a German-American
historian, philosopher and political theorist.*

THE DISCUSSION of the whole problem of technology, that is, of the transformation of life and world through the introduction of the machine, has been strangely led astray through an all-too-exclusive concentration upon the service or disservice the machines render to men. The assumption here is that every tool and implement is primarily designed to make human life easier and human labour less painful. Their instrumentality is understood exclusively in this anthropocentric sense. But the instrumentality of tools and implements is much more closely related to the object it is designed to produce and their sheer 'human value' is restricted to the use the *animal laborans* makes of them. In other words, *homo faber*, the toolmaker, invented tools and implements in order to erect a world, not – at least, not primarily – to help the human life process. The question therefore is not so much whether we are the masters or the slaves of our machines, but whether machines still serve the world and its things or if, on the contrary, they and the automatic motion of their processes have begun to rule and even destroy the world and things.

Hannah Arendt, *The Human Condition* (University of Chicago, 1998), 151.

Hollie Chastain, *March*, 2020.

## A Lindisfarne Cross

It begins with what is broken,
cast off, abandoned, lost beyond all usefulness or artifice.
The abyss leaves only a remnant on the beach:
a gobbet of green glass,
two twigs,
a skein of fishing line,
the last light ring of a worn-out shell,
three tiny beads.
These withered, weathered, weakened pieces are chosen,
placed with tender delicacy, fresh-bound to make
a coracle, tossed upon the ocean, almost overwhelmed
yet crowned by a bright halo,
tucked under the nook of a cross.
This is a reckless act.
Strange, that what is most frail is what endures,
leavened into iridescent beauty
in the work and cost of love,
bearing with the depths, willing
a new word to be spoken,
and what is done and made with pain.
It begins with what is broken.

*MICHAEL MANNING*

# ChatGPT Goes to Church

*Should large language models write sermons and prayers?*

**ARLIE COLES**

I N A 1999 EPISODE of the church comedy *The Vicar of Dibley*, the PCC (Parochial Church Council) meets to discuss how to lift vicar Geraldine's spirits after her breakup with the womaniser Simon. Alice, the daft though creative verger, has a solution. 'You know the series *Walking with Dinosaurs*?' she ventures. 'Well, they recreated the dinosaurs digitally, just using a computer. I thought maybe we could do the same with Uncle Simon.' David Horton, the churchwarden, repeats dryly: 'Recreate him digitally?' 'That's right,' says Alice, 'then send the digital Simon round to the vicarage.' A beat passes, and Mr Horton clarifies: 'So we get a holographic, two-dimensional human to marry the vicar?' Alice nods, and Mr Horton looks around for help responding to her technically impossible and morally absurd suggestion. 'Does anyone

spot the defect in this plan?' he asks. No one does, and the PCC votes through Alice's motion.

Working in artificial intelligence research since ChatGPT launched often makes me feel like hapless David Horton. We researchers are surrounded by onlookers and their suggestions, some having neither desirable goals nor methods in the realms of reality. For example, some suggest outsourcing middle and high school teaching to a chatbot. Its inaccuracies 'can be easily improved', claims an academic dean at the Rochester Institute of Technology. 'You just need to train the ChatGPT.' Or, as the tech entrepreneur Greg Isenberg suggested last year, we could task a language model (LM) with writing and marketing the next Great American Novel; all we have to do is code up a programme and 'start selling'. Each time a public figure urges this kind of unfettered and unrealistic application of LM technology to tasks far too human to morally bear automation, I hear the harried churchwarden's voice: *Recreate him digitally? Does anyone spot the defect in this plan?*

Yet in many cases the PCC has voted the dubious motions through. Modern LMs have launched a thousand bullish startups and a thousand uneasy think pieces. Many doubt the wisdom of hastily applying LM technology to areas classically sitting at the core of human creative activity – writing, teaching, interpreting – especially as the general public discovers what machine learning researchers already know: LMs are not omniscient and can, in fact, generate nonsense. Moreover, some worry, if LMs are by nature rubbish-generating machines, does using them relegate us to mediocrity? Will we let our human creativity atrophy? And, seriously, what real need – not simply the desire for the last scrap of profit at the expense of human pursuits – do LMs give an answer to?

Why have LMs suddenly and dramatically seized our attention? Where should the church say, 'Thus far and no further'? Few are truly equipped to assess the deadlock, but as a speech and language AI researcher and a churchwoman, I will attempt the task.

'Father Justin', an AI chatbot designed to answer questions regarding the Catholic faith, started giving false guidance and even offering one user absolution as a 'real' priest, raising questions over the limits of AI in the church setting.

T HE PUBLIC REACTION to ChatGPT's launch in November 2022 exceeded industry expectations. OpenAI, its creator company, had called the launch a 'low-key research preview', and many in the wider research community were poised to see it as another incremental though impressive improvement in the long line of LM research. LMs are a workhorse class of statistical model that, as the public now perhaps knows, have one job: to return the next most probable item in a sequence. An LM will predict 'mat' given 'the cat sat on the' because, roughly, we give it a count of how many times 'mat' appeared in that context in some other text corpus, and if that count is high enough, we set it to output that item given the context. LMs have powered common AI applications for decades, including the predictive text suggestions in your next tweet and your last email, so why the explosion of public interest now?

Today's chatbots offer the glittering peril of instant gratification. Products like ChatGPT are strictly speaking not just LMs, but LMs *plus* extra statistical steering to facilitate a question-and-answer format, *plus* a slick web interface enabling any person on earth to type in a query and receive a fast response. This chatbot framing is technically incidental. But it is rhetorically and psychologically powerful, a rapid feedback loop so easy to enter that it has implicitly taught the public that the purpose of an LM is to generate content on demand.

And to generate answers. Our tendency to conflate *quick* responses with *correct* responses when talking to humans also transfers onto the chatbot, which we can't help but personify. A fast and confident-sounding chatbot may mimic the authoritative voice of a reference work, and it may draw from and contribute to our habits of laziness when seeking out truth and our impatience when engaging each other.

Researchers know – but seldom effectively communicate – that indispensable LM applications sit just one level deeper than the 'type query, get content' chatbot paradigm. That's because LMs are only *accidentally* content machines; they are *substantially* a dense statistical representation of relationships between words. Eliciting those relationships for a downstream task can be valuable. Borrowing heavily from John Firth's distributional semantics maxim, 'You shall know a word by the company it keeps', LMs compress and store information about the whole distribution of words in all the text they see. Their insides and outputs are a mathematical study of language as it is actually used, and researchers can and do leverage that information not to drown us with spam, but to promote our good.

One can use a printing press to print libel. But that is not what it is for. What, then, ought LMs to be for?

They are, already, part of innumerable systems that humanise, not alienate – speech recognisers whose generated transcripts are important accessibility tools for deaf people; translators that allow immigrants in tight spots to communicate; record systems that alleviate medical professionals' documentation burdens and allow them to spend more time at the bedside. Their value can even return to the fields that provided their data. LMs are helping to decipher dead languages, restore lost ancient inscriptions, and predict protein structures. They are tools that, in the hands of intrepid researchers and, yes, entrepreneurs, are well suited to facilitate our exploration of the world and our rapprochement with one another. But none of these mentioned applications use LMs as cheap content machines; they are harder to understand (and get

*Arlie Coles is an AI researcher, linguist, and software developer specialising in deep learning for automatic speech recognition and natural language processing. She has a master's degree in computer science from the University of Montreal.*

less press) than the instant feedback an impressive or appalling chatbot provides. Our attention spans are short; our demand for content is high. The high-strung discourse around LMs is, in a sense, what we deserve.

AND WHAT OF THE CHURCH? Some clerics have voiced concerns about AI writ large, ranging from the pastoral – how should a priest help a parishioner following an automation layoff? – to the theological – can an LM be possessed by a demon? (And, if so, is it the same demon that always pops up in the printer when it's time to produce bulletins during Holy Week?) Others are more optimistic about the possible elimination of drudgery. This divide mirrors the one in the public's discussion of AI, with one faction wishing to seize the low-hanging fruit, while the other asserts that from the beginning in the Garden, knowledge of when to appropriately seize fruit has hardly been humanity's strong point. The fact is, if the church implements suggestions for how to use LMs that are as shallow and dehumanising as the suggestions that have lately come out of secular society, we will live to regret it.

The first truly interesting suggestion for LM use in the church is to leverage them for organisation of language-based materials. Many helpful assistive technologies flow from this: subtitling, transcription, translation of services, searching past sermons or resource documents, and so on. Insofar as these ideas increase access to the life of the church, they can be healthily pursued. But two dangers lurk around this corner.

One temptation is to jump from using LMs to *organise* text to using LMs to *interpret* text – including the ultimate text, the Bible, which ought to be wrestled with and interpreted in the human community knit together by the power of the Holy Ghost. Some enthusiasts have advised pastors to save time by using LMs to generate devotionals and Bible discussion questions, but one cannot excise the humanity from these and still obey the call to personal and communal wrestling that Holy Scripture demands.

Technically, the returns will diminish because of the nature of LMs: they will return shallow text probabilistically biased toward any religious text in their training corpus. Spiritually, while LMs may marshal text effectively, they can neither 'read, mark, learn,' nor 'inwardly digest' it. Meditating on divine words is what human beings do in their inner being. This technically cannot and morally should not be automated. Mary could not have outsourced her pondering of the angel's words to an LM, not only because an LM's next-item-prediction objective is not pondering, but also because it would have denied those words' ability to form *her*. A pastor might provide his congregation with a Bible study written by another pastor or a church father – but its author is still a person in relationship with the universal church, an inward digester, who, though having died, is alive in Christ and truly helps form the congregation. Rejecting this in favour of artificially generated text is an affront to the reality of the communion of saints.

Some have encouraged training LM-based chatbots on the Bible; others, while warm to the idea, have exhorted machine-learning practitioners to erect guardrails to ensure these LMs return text congruent with both the Bible and their users' theological stances. As one such practitioner, let me say clearly: there is no way to guarantee this. Because LMs are not comprised of retrievable data and hand-coded interpretable rules, but rather abstracted statistical reflections of their training data, perfectly imposing such guardrails is an unsolved problem to which there may be no final answer. LMs do not look up information. That is not how they work. This makes LMs a fundamentally inappropriate tool for handling the Bible, where information retrieval accuracy and interpretive fidelity are nonnegotiable. Engineers know that building a bridge with the wrong material will cause it to fall down, and good engineers refuse to build bad bridges; let the reader understand.

Another temptation is to slip from using LMs to organise church content to using them to commodify that content. In our postpandemic era of broadcasting everything online, the tendency to turn acts of worship into acts of marketing is hard to resist, and sermons are a facile target for this trap. Outsourcing follows commodification, which could here result in an outright denial of the duty to preach. Some leaders in my own Episcopal Church are already situated at this dangerous pass, placing sermons in the same category as parish announcements – items whose automation will free up overladen clergy for, presumably, *real* pastoral work.

Karl Barth's notion of preaching as an exposition of the Word of God is a helpful counterweight to these plans for LM-generated sermons. When a sermon is planned by a minister and proclaimed to the people through the mouth of the church, the Holy Ghost assists the delivery and makes it the very Word of God to the hearer. Not all denominations will agree on this semi-sacramental view of preaching, but all should agree that sampling from a next-word predictor is an inappropriate and unethical replacement for it. A pastor is responsible for the congregation's spiritual formation, to which preaching is central; who could delegate this to a synthetic-text machine? Accidentally generated heresy is a technical failure; a pastor refusing to speak from the heart and preferring to generate the most probable word sequences for a sermon to the congregation in his care is a moral failure.

The final stop on this dubious trajectory is saddling LMs with the task of liturgical composition. Congregants at a Bavarian church that attempted this found the service trite and unsettling, some even refusing to join in saying the Lord's Prayer. Their discomfort was well founded: this type of LM use encroaches on the unique vocation of humans within the whole creation's worship of God and creates a liturgical absurdity that we feel in our gut.

All of creation expresses a cacophony of praise to its Creator. 'One day telleth another' of God's glory, says the psalmist, where 'there is neither speech nor language; but their voices are heard among them' (Ps 19:2–3). In Isaiah, we hear that 'the mountains and the hills shall break forth before you into singing, and all the trees of the field shall clap their hands' (Isa 55:12). Yet God appoints one

Some funeral homes in Japan use androids for chanting sutras over the deceased and livestreaming services to distant family members.

creature to collect all the noisy voices of creation and consolidate them into ordered expression: the human being, whom God endowed with the richest linguistic faculties. Language powers are integral to our being made in the image of God. Through them, we are able to rationally organise and sit in dominion over creation and cultivate it, enacting God's goodness to it and bringing forth the harvest of its praises to offer them to God. 'O *all ye Works* of the Lord, bless ye the Lord,' the Prayer Book canticle has us cry before enumerating these Works, from the lightning and clouds to the whales and all that move in the waters. 'Praise him, and magnify him forever!'

Of all creation, the human is the priest, mediating between it and God, in part by the ordering power of language. This priesthood is of all believers, as language faculties are universally inherent in us *in potentia* (and in actuality, far beyond what we might think; indeed, deaf babies babble in a structured manner with their hands, and signed languages possess full phonological and syntactical systems).

When we shirk our duty to use our language faculties to worship God, the infraction is multiple: we not only fail to offer our own sacrifice of praise and reject God who would respond in goodness, but we also deprive all created things from joining their natural expressions of praise to his. 'He that to praise and laud thee doth refrain, / Doth not refrain unto himself alone,' warns George Herbert, 'But robs a thousand who would praise thee fain, / And doth commit a world of sinne in one'. The poet has the obligation to sing; the poet also needs to sing for his own sake because his song, reaching out to God, changes him. Given this, the idea of liturgists handing their jobs to LMs is farcical, as laughable as a man whose daughter needs surgery sending a calculator to be operated on instead. We present a dumb machine in our place, hiding from God who hears us when we call and transforms us when we ask, and doing collateral damage to other creatures in our care.

Christians must instead look to Jesus. The Word of God, coming from the mouth of the Most High, became flesh. He took on materiality and carried it to the right hand of the Father at his ascension. He is the great high priest who mediates between God and creation, through whom all creation will be redeemed on the last day. He shares our humanity, collecting the noisy and often incredibly sideways praises of our human life in himself and ordering them according to the divinity of the Logos – and he calls Christians to follow him in this. We, therefore, cannot abandon our role in orchestrating, via our own language, worship in which all creation participates. Outsourcing this to a text generator is absurd in the extreme, a near-literal abdication of the throne God set up for human beings, who, while made a little lower than the angels, have everything in subjection under their feet through Jesus, the eternal Word, true man yet very God.

WHERE DOES THIS leave the church? The fixation on LMs as content generators, tools that circumvent the necessity of thinking together, is symptomatic of a deeper disease, developing out of our failure to integrate our unprecedented technological interconnectedness with the bodily realities that true Christian – true human – interdependence demands. The church uncritically glomming onto the latest LM for its liturgical, educational, or pastoral work will compound the harm in this area already inflicted by the long, lonely slog of the pandemic.

There is no world where deferring preaching and pastoral care to a text generator does not end with deterioration – first of formation, then of the clergy, and finally of the people in their care. As more seminaries move online or shutter altogether, and more clerics are forced to work full-time jobs at part-time pay, what else can the replacement of their functions by LMs spell for those in pastoral need?

There is also no world where increased comfort with liturgical automation does not end with attempts to obviate the sacraments. Our Lord peskily attached himself to material things that force Christians to keep one foot in reality. But

this tether is threatened, not by him but us, and not to his detriment but ours, if we go down the path of thinking that a machine can compose or recite a prayer to almighty God.

Meanwhile, captive to the view that sees LMs as content machines only, people who rightly object to this direction in the church will retrench, potentially causing a churchwide neglect of the opportunities to use machine learning well – opportunities less flashy but more helpful. Great promise exists for LMs in service of better research, the offloading of true drudgery, and increased access to various aspects of public and personal life for the linguistically barred – and indeed LMs have been fuelling all these things without public fanfare or objection for many years. Opposition will be created in areas where none need exist.

A renewed belief in the communion of saints is a necessary part of the treatment. Saint Paul says that every member of this body is needed. All contribute something irreplaceable by anything else animate or inanimate, carbon or silicon. The incorporation of Christians as human persons into one body, that of the divine Word himself, is a profound mystery that cannot in fact be menaced or usurped by a text generator, try though we might through active promotion or doomsaying alike.

We should take heart in that, and then take up and read – and write, communicate, and contemplate, first enjoying and maintaining those gifts from God without fear of their replacement. Then, having received freely, we should freely give. Using language technology for the right purposes will facilitate the exercise of those gifts by those who would usually be restricted from their use by physical condition or temporal station. Neither slick demos nor technical party tricks can get us there. What is required is nothing less than a true love of God and neighbour, which no machine can generate. ➤

# Taming Tech

*How the Bruderhof community tries to be intentional
about personal technology.*

**ANDREW ZIMMERMAN**

THE BASKET QUICKLY became known as
the 'handy garage.' (*Handy* is German
for mobile phone.) In reality it wasn't
particularly handy when none of us could turn to
Google or Wikipedia to settle arguments at the
dinner table. For a few days, I noticed everyone's

discomfort when a notification sounded from
across the room and we each strained our ears to
discern if it had been ours. Some days at first, one
of us would notice that the basket looked strangely
empty and would pass it around the table again to
retrieve a few forgotten phones. But soon enough,

Pedrita, *Untitled*, reclaimed tile trims, 2018.

we all habitually parked them in the garage upon entering, and it did seem that conversations lasted longer and had more substance.

Here at the Gutshof, one of about 25 Bruderhof communities inspired by the first gathering of believers described in the Acts of the Apostles, we have no private property but share a common purse and common life. Some Bruderhofs are home to several hundred people, while others are smaller households like ours, but in all cases, we share work, meals and worship. Members cook food, or grow the food, or wash the laundry, or mow the lawns, or work in the furniture factory – and draw no salary for doing so. At the larger communities, we operate primary schools, middle schools and high schools for children of members as well as neighbours. Onsite clinics provide free medical and dental primary care. Elderly members are offered ways to meaningfully participate in the work until they're no longer able and are then cared for within the community.

Jesus tells his disciples to remain 'in the world but not of the world' (John 17). As part of our collective effort to discern what obedience to this command might mean today, we adopt an attitude of caution towards new technologies, just as we take a countercultural stance towards other norms of contemporary society, for instance money (we hold no private bank accounts or credit cards), career (we work where we're asked to, not where we want), fashion (simpler is better), and sex (chastity and faithfulness in marriage).

Such caution, when directed at newer technology such as smartphones, does not necessarily mean hostility. And it's not a blanket ban. We consider that tools, whether simple or high tech, are used to build up the community, advance our mission as a church, and help individuals and families to

**About the artwork:** Inspired by a set of lost (and found) photographs, *Lost and Found*, an art series by Pedrita, is a contemplation on the loss and recovery of our individual memories, our collective referents, and our cultural heritage.

Spanning the analogue and the digital world, the past and the present, the artworks were created from discontinued industrial tiles arranged to reflect their original composition of pixels or photographic grain.

Pedrita is a Lisbon-based design studio founded in 2005 by Pedro Ferreira (b 1978) and Rita João (b 1978), drawing inspiration from traditional Portuguese forms and techniques.

flourish: we have long used electricity, cars, robotic welders, automated logistics systems and medical technology such as diabetes monitors.

Despite this openness to positive uses of technology, we've also drawn a number of boundaries around technology we choose *not* to use; these can strike newcomers as fairly strict. These guidelines aren't moral judgments – we won't tell you you're a bad person if you do these things – but arise out of a communal sense that certain technologies prevent flourishing more than they help. When television came along, we chose not to have TVs in our family apartments. Members typically don't use social media, except for business or

---

*Andrew Zimmerman lives at the Gutshof, a Bruderhof in Austria, with his wife and three children.*

outreach purposes (for example, to promote this publication). Our children don't get smartphones until they finish secondary school. Having chosen a life of voluntary poverty – in the sense of having no personal spending money and wanting to live simply – most community members are little affected by the technological consumerism that looms large in society, from online advertising to gambling sites. It's hard to make impulsive Amazon purchases when you don't have your own money. For similar reasons, we also generally aren't early adopters of personal tech, since we don't choose our own devices (so no chance of picking up that latest Apple gadget unless you can make a case to your fellow members that you actually need it).

Even where we welcome technology for its efficiency, we've learned to appreciate that it can be healthy to give it a rest now and then. We may use automatic dishwashers after a communal meal for two to three hundred people, but we'll also fill up a large sink, surround it with four or five men and tackle the pitchers and serving bowls. (There are memorable conversations to be had while scrubbing out pans.) Other times we'll get out shovels and wheelbarrows for landscaping rather than using a tractor, for no other reason than that it's good to get your hands dirty and your muscles tired while working together with others.

The Siren in our pockets was adversely affecting many forms of creative contribution to community and society.

W E THOUGHT we had a pretty grounded and effective approach – until the pandemic struck. Like many churches, we rolled out infrastructure and devices during the various lockdowns of 2020 and 2021 so church services and members' meetings could be held online and so that many of us could work from home. The Bruderhof's secondary schools issued tablets to accommodate virtual learning. Prior to that, few residential apartments had Wi-Fi and less than a quarter of adults had smartphones. Of course, the technology had upsides – it enabled us to keep connected, united in prayer and mutual support, through some difficult months.

But as the pandemic receded and in-person meals, meetings and work resumed, the phones, tablets, laptops and Wi-Fi stayed.

How much the ground shifted during those trying months only became clear a couple of years later. Within our close-knit communities, we began to notice that it seemed to be more difficult to hold conversations without interruption. Communal meals and meetings were interrupted by unsilenced phones. Work-life balance seemed out of whack in many families, with more parents working from home at odder hours. More personal technology was available to children and families found themselves watching more movies and sports and playing fewer board games and less pitch-and-catch. Although the beginning of the pandemic had brought a renewed interest in crafts, hobbies, baking and music, those heady DIY vibes seemed to have faded, a worrying trend in a church community that has always held practical creativity in high regard.

We realised that we were at risk of losing a valuable portion of the face-to-face interactions that are essential to the Bruderhof's way of life. We were reminded that tech is not in fact neutral: It can be a passive hindrance to community by stealing our time. And it can also be an actively destructive force – not just among teenagers, but among people of any age.

This, of course, is not a problem unique to the Bruderhof. Over the past decade, research featured by writers including Jean Twenge, Nicholas Carr, Cal Newport and most recently Jonathan Haidt has shown the grave costs of the introduction of smartphones and social media, and these costs appear to be highest for children and adolescents. Haidt's book *The Anxious Generation* (March 2024) lays out starkly troubling trends: drastic increases in anxiety, depression and suicide. 'By a variety of measures and in a variety of countries, the members of Generation Z (born in and after 1996) are suffering from anxiety, depression, self-harm and related disorders at levels higher than any other generation for which we have data,' he notes. We were simply seeing the evidence of what such authors have long argued: phones are bad for our mental and physical health. The Siren in our pockets was adversely affecting many forms of creative contribution to community and society. In addition, more members found themselves struggling with temptations to pornography and excessive gaming, which are more accessible through personal tech.

Many of us sensed that as late adopters we were actually worse than some others at setting boundaries for tech use and realised that we needed to put the brakes on.

There was broad (though not universal) consensus in our communities that the growth of personal tech had become a serious problem.

Pedrita, *Untitled*, reclaimed tile trims, 2018.

According to the rule of our common life, it was up to the body of members to find a solution. One such possible solution, which some were eager for, would have been to develop strict new rules to apply to each of our communities to combat perceived misuses of technology. Since we make decisions unanimously, this would have meant members agreeing on limitations that everyone would be expected to accept and uphold.

We chose not to take this path, which would inevitably involve a degree of legalism and rigidity. It would not be in keeping with our identity as a community founded on trusting and honest relationships rather than enforced conformity. Instead, a team of pastors (including myself) and IT professionals assembled to address issues and bring forward solutions. We went in knowing we could not roll back the clock, but we were willing to see how we could become more disciplined and intentional about our use of personal technology.

In turn, we reported to our congregations about the problems we saw and asked them too to have discussions and help come up with practical suggestions. In these meetings, rather than focusing on the 'evils' of smartphones and social media, we looked for specific, achievable solutions to problems like a dearth of connection or a rise

in loneliness. Members came up with guidelines for appropriate phone use in various communal settings, to reestablish some basic social norms. Our emphasis was not on controlling each other but on mutual accountability, on nurturing the positive vision of what community is for and letting personal tech find its place within that.

The proposed guidelines, which have been adopted with varying degrees of enthusiasm, included suggestions around how and when earbuds should be worn in communal workplaces such as workshops, kitchens, or gardens so that real conversations could flourish instead of everyone listening to their podcast of choice and the expectation that we wouldn't check our phones during communal gatherings. And while it need not be strict, we realised we would all benefit from some form of Sabbath observance, where it's expected

Pedrita, *Promenade*, reclaimed tile trims, 2018.

you *won't* answer calls, texts, or emails on Sunday.

Some things, however, were stricter: we reaffirmed our stance on not providing smartphones for kids before the end of secondary school and on avoiding personal social media use. In cases where a social media app might be necessary, the users have to make their case: for example, some of our college students might need a certain app to receive their homework assignments. A sports team may arrange practice on a social media page; artists, photographers and potters may use Instagram to showcase their work. Several community schools decided that teachers being on their phones was an issue of child safety, so they now have staff turn off their phones while teaching. By no means are we the only ones to do this. I recently visited a church in Spain where the children's room had a prominent sign on the door saying: 'Absolutely no use of your mobile device when caring for our children!'

Our team also educated ourselves more fully on dangerous aspects of technology, such as the effect of social media on teens' mental well-being and made ourselves available as a resource to parents who might wish to have these discussions with their school- or university-age children but feel unequipped. And for young adults, we renewed our emphasis on coaching them to learn to use tech well, since it's inevitable that they'll be using it. As with much of life, our attitude to personal technology requires a balancing act; like so many people these days, we are trying to find the right place for it. Our conversations and adjustments to tech use are ongoing.

We continue to ask ourselves what it means to be countercultural. From the early church on, followers of Christ have been out of step with the world; in fact, the gospel demands it. Yes, it can be hard, but maybe that's the point. 'We live amid a carnival of distractions and we have to daily wage war on the sin in our hearts that wants to yank us away from the kingdom purposes we find in our faith,' writes Chris Martin in his excellent book

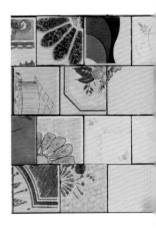

> We realise better today than we did two years ago that we should be too busy caring for our neighbours to spend excessive time online.

*The Wolf in Their Pockets.* 'This is why we need the community of brothers and sisters in Christ, to help us walk the narrow path when we'd rather just go our own way.'

As often as we stumble off the narrow path, we try to stumble back on again. To return to the 'handy garage' experiment that emerged in my small community: as far as I know it has not been adopted by any other Bruderhof, but because it arose as a result of our discussions here, it was meaningful to us. It changed our habits enough that we were able to put the literal parking of the phones on pause, for now.

As new developments in tech continue to present new challenges, we're committed to finding a healthy approach together. Most of us realise better today than we did two years ago that we should be too busy caring for our neighbours to spend excessive time online.

'Do not be conformed to this world,' Paul tells the Romans – and us – 'but be transformed by the renewal of your mind'. The distractions of viral trends, news alerts and targeted advertising tend to conform us to what the rest of society wants. Transforming our minds is a lot harder. And even legitimate work-related tech use can pull our focus away from our brothers and sisters, and from God. But together we can overcome this pull and I believe we will. ⤳

Du Zhenjun, *Tower of Babel: Crusade*, photographic collage, 2011

# Building Babel

*Our struggle with technology starts in Genesis.*

**ALASTAIR ROBERTS**

**M**AKE YOURSELF SAFE. Make yourself a name. Make yourself eternal.
These three are the drivers behind much of what we humans do, according to the story that the Hebrew Bible tells about the development of human civilisation. These are what we, as a species, have treated as our telos, our purpose.

They are not what God has made us for. But the tricky thing is that they're not entirely unrelated either. Here's what God said about our purpose, when he first made us:

> Let us make man in our image, after our likeness. And let them have dominion over the fish of the sea and over the birds of the heavens and over the livestock and over all the earth and over every creeping thing that creeps on the earth.

So, God created man in his own image, in the image of God he created him; male and female he created them. And God blessed them. And God said to them, 'Be fruitful and multiply and fill the earth and subdue it and have dominion over the fish of the sea and over the birds of the heavens and over every living thing that moves on the earth (Gen 1:26–28).

The human project goes somewhat off the rails with the Fall, when Adam and Eve aim to set up an independent judiciary, 'having dominion' in the sense of taking the judgement seat before they are raised in the fullness of time to the divine council, judging on their own terms. This judicial autonomy is only the first example of humans taking their good human task, the creation mandate, and attempting to carry it out without reference to the God who gave them that task.

The story of building civilisation under the conditions of the Fall is a story of technology – language and politics wrenched out of the order they were meant to serve, used to serve grasped glory rather than the glory that is given. This pattern emerges very quickly after the second generation of humans is born.

After being exiled from the presence of the Lord for the murder of his brother, Cain fathers a son, Enoch, and founds a city named after him. The city is a manmade system that he inhabits, a world in which he lives not as a king who is the living icon of God, but as a king under no other authority. Cain's city, as Jacques Ellul argues in *The Meaning of the City* (1951), is an act of piracy, an attempt to harness the creation to his rebellious will and inaugurate a world of his own in opposition to God's Eden. In naming his son and city, Cain attempts to secure a legacy for himself in the teeth of death and divine judgement. His descendants are responsible for the first great technological innovations in the biblical record: the invention of musical instruments and metalwork.

After the flood, the city resurfaces in Genesis 10 with the figure of the great champion and empire-builder, Nimrod. A descendant of Noah's rebellious son, Ham, Nimrod pioneered the Mesopotamian vision of kingship. Although Nimrod's biblically recorded acts are those of city-founding, his reputation is recorded as that of a hunter, so great as to be referred to as a 'mighty hunter before the Lord' (Gen 10:9). Associations between hunting and kingship are ancient, cross-cultural and persistent. The hunter embodies the increased dominion granted to Noah after the flood, and the king represents its extension into the power of death and conquest over his fellow men. Again, the pattern: God has given Noah the right to hunt, to eat meat. Nimrod, though, takes that gift and turns it into something that boxes God out of the equation.

And the civilisation Nimrod ends up building is the emblem of the new militarised state-societies of the third millennium BC. In hunting, strength and skill in arms could be honed, proved and displayed. Bands of hunters could develop into military hierarchies and, even when kings no longer led their men in battle, a king's devotion to the hunt was a vital metaphor for the meaning of kingship. The prominence given to Nimrod's

*Alastair Roberts teaches for both the Theopolis Institute and the Davenant Institute. He and his wife,* Plough *editor Susannah Black Roberts, split their time between New York City and the United Kingdom.*

hunting prowess might suggest that subjugation was central to the logic of his kingship. The connotations of 'before the Lord' are debated by commentators, but the contextual setting seems to favour negative ones. Nimrod, whose name seems to be a play on the Hebrew root for 'rebel,' recalls

nothing that they propose to do will now be impossible for them' (v 6).

This judgement upon the Babel builders, upon those whose hearts are turned backwards, is reminiscent of the Lord's judgement upon man in Genesis 3:22–24. There the fact that people had

---

A naive reading of Genesis might give the impression that the scattering of the Babel builders is motivated by the Lord's concern about a threat to his own throne. Yet the threat is primarily to human beings themselves.

---

the Nephilim and the mighty 'men of name' who arose from them (Gen 6:4), becoming, in his power, like a god among men.

Among Nimrod's great acts was the founding of the city of Babel, or Babylon. Besides the similarity between the origins of Nimrod's empire and the description of the origins of Israel's bondage in Egypt (Exod 1:8–14), consideration of the nature of such enterprises would suggest that Nimrod's vast city-building ambitions must have rested upon the shoulders of innumerable slaves.

Attentive readers of the account of Genesis 11, however, might notice that the Babel builders begin not with a plan to build a city and tower, but with the discovery of a technique for firing bricks. The determination to build the city and tower seemingly arises, at least in part, out of humanity's intoxication with new technological potential. As in the case of the exiled Cain, a concern to secure their legacy through the achievement of renown motivates the builders. Their ambition has both a horizontal and a vertical impulse: to build a city that gathers humanity under Nimrod's sway and an immense tower that represents the godlike greatness of their name.

The project of Babel is frustrated by the Lord, but not without an admission of the genuine danger that the city and tower represent: 'This is only the beginning of what they will do. And

prematurely seized the knowledge of good and evil for themselves, becoming like one of the gods – but more lonely, a divine council all to themselves, led to the Lord casting Adam and Eve out of Eden. Obtained in such a rebellious fashion, the knowledge of good and evil, the wisdom associated with rule and judicial authority, manifests the human aspirations for autonomy and for enjoying the role of God in our own world. Indeed, the Lord grants that human beings achieved something of his intent when he declares that the man had 'become like one of us' (Gen 3:22), like one of the heavenly rulers. To deny this rebellious ambition full rein, the Lord cuts the man and woman off from the Tree of Life and its promise of immortality.

Humanity's hubristic quest for autonomy does not cease, however and with innovations in technology, as Babel demonstrates, it is only inspired to reach for greater heights. In the city of Babel, we see the rebellion of our first parents in the garden coming of age, the 'beginning' of a new phase.

The Lord's frustration of the Babel project provides the backdrop for the narrative of the call of Abram and the Lord's assurance that he would make Abram's name great (Gen. 12:1–3) – and that therefore Abram did not need to do one of the Ancient Near East's 'make my own name great' projects. He could leave the city and venture out. And in the story of Abraham's descendant, Jacob, we have a recollection of the story of Babel in a ladder that reaches to heaven and the play upon the meaning of Babel ('gate of God') in the patriarch's awed response: 'This is none other than the house of God and this is the gate of heaven' (Gen 28:17).

By the end of the seventh century BC, the Abrahamic alternative to Babel seemed to have failed, though, with Jewish exiles being taken in captivity back to the land of Babylon. The story of Daniel begins with this tragic return of Abraham's offspring to the land of his origin. However, the careful reader of the Book of Daniel might notice familiar themes pervading it, variations of the ancient story of Babel.

The Book of Daniel is a book about the distant heirs of the Babel builders, of kings with ambitions for godhood, of attempts to bring all peoples under a single human power, of the frustration of language and interpretation, and of the downfall of great towers. In three successive chapters, the book describes three towering structures: the great statue in Nebuchadnezzar's dream (ch. 2), the golden statue of Nebuchadnezzar himself (ch. 3), and the tree with the top reaching to heaven (ch. 4). All three represent the empire's attempt to subdue all things under it.

The first statue symbolises a succession of empires, beginning with the golden head of Babylon. This massive image, a marvel of human creation and engineering, its different materials representing the amalgamation of all peoples through human power, is brought down and destroyed by a divine stone cut without hands. The events of chapter three might be understood as Nebuchadnezzar's resistant response to his dream, as he sought to bring all 'peoples, nations and languages' together through the collective worship of a gigantic golden image erected on a plain. This is a deliberate attempt to reverse the Lord's judgement of Babel, which had scattered these peoples and languages. The fiery furnace into which the faithful Hebrews were cast was likely used in construction of the image, illustrative of empires' attempts to melt peoples down

into manifestations of their absolute sovereignty. In chapter four, the tree with its top reaching to heaven is Nebuchadnezzar himself, his kingdom gathering the peoples under its shade and in its branches. In a manner that is once again reminiscent of Babel, a 'holy one' descends and chops down the tree, scattering those who had dwelt under it. The proud king Nebuchadnezzar, who had pretensions to the throne of God, was humbled, being reduced to a bestial state. One is reminded of Aristotle: 'He who is unable to live in society, or who has no need because he is sufficient for himself, must be either a beast or a god'. Nebuchadnezzar, refusing to live in a polity, in brotherhood and seeking instead to found an empire, refuses to be a man and instead seeks to be a god. And God makes him a beast.

Along with the reminders of Babel's tower, Daniel is a book of confused language and failures of interpretation. None of the king's wise men can tell him his dream or interpret it in chapter two, and they fail again in chapter four. King Belshazzar's and Babylon's doom is foretold by the mysterious writing on the wall in chapter five. Besides being an object of interpretation, language is also a means of human agency and rule. Like the Lord, we seek to make our world by the power of our word. Yet the recourse to the more universal language of music to bring all peoples together fails in chapter three, and both Nebuchadnezzar and Darius find themselves trapped in their own words.

Babel illustrates the technology-assisted ambitions of people for dominance over others. A naive reading of Genesis might give the impression that the exile from Eden and the scattering and confusion of the language of the Babel builders is motivated by the Lord's concern about a threat to his own throne. Yet the threat is primarily to human beings themselves. People make dangerous gods for others – and for themselves – and as they progress beyond the infancy of the garden, as the Mesopotamian god-kings arise, along with the

Du Zhenjun, *Tower of Babel: the Wind*, photographic collage, 2010

later vast empires of Babylon and its successors, this truth becomes increasingly evident. Those who seek to usurp God's rule establish houses of bondage for their fellows, but they also lose grip on their own humanity.

Daniel presents us with a further aspect of this when the word through which the king seeks to control his world turns against him. In chapter six, Darius's officials use the king's word of law to attempt an internal coup so they can remove Daniel, of whom they are jealous. The irrevocable character of the king's law, which should represent his sovereignty, turns into a rod for his own back. Something similar happens in chapter three, where

malicious officials use the king's word to advance their own petty court rivalries, making him cast the Hebrews into the fiery furnace. It is 'the law of the Medes and the Persians, which cannot be revoked' (6:8), which subverts the king's own desire to do justice.

The capacity of the word to turn against its supposed master is a biblical warning against the dangers of humanity's overweening ambitions for autonomous dominion. Unlike the divine Word, who is always with the Father and in whom the Father always dwells, the 'words' by which people seek to fashion and control their worlds can escape and betray them. Seen biblically, the role of legal and technological structures in the rebellious systems that people invent – parodies of the creation mandate – should caution us: the words and works of creatures striving for a rebellious autonomy have their own tendencies to autonomy and can imprison those who supposed themselves to be their masters. Make an idol – political or technological – and it will turn on you. The Lord's frustration of humanity's ambitions through death, the resistance of creation, the confusion of language, the scattering of humanity and other such means are ways in which he saves us from those who seek to be, or create, gods.

Where does that leave us? What can we learn from the projects of these god-kings?

It is easy to labour under the illusion that things like the car, the internet, or our economic systems more generally are creatures of man, under human control. Yet it should not take much reflection to appreciate the degree to which such technologies, techniques, processes and entities have autonomous logics of their own, logics that can exert a godlike influence over their former creators. Mammon, as Jesus observed, is something that people serve. It does not need to be a self-conscious artificial superintelligence to hold sway over people's worlds and hearts every bit as strongly as did the ancient gods of the pagans.

The more those carrying out the rebellious version of the creation mandate imagine that they are extending their powers, the more they can fall under the thrall of their creations.

We tend to refer to the totality of our realms of human interaction and cohabitation in ways that foreground their human character, yet such ways of speaking can blind us to the degree to which our Babelic ambitions have crafted vast impersonal technological systems within which all of us, even those supposedly wielding power, are captive and to which we are subservient. Like Nebuchadnezzar, proud in our technological capacities, our humanity is debased as technology empowers our passions to dominate us. Seeking to exercise our good human abilities in illicit ways to attain godlike power, we lose not only the godlike power we thought we had, but even our good human abilities themselves.

Babel, where we first see the power of technology serving humanity's quest for godlike autonomy, is a symbol of the City of Man and its apocalyptic downfall. Its biblical antithesis is not merely the Garden, but the glorified garden city of the New Jerusalem. The alternative to the hubris and rebellion of the technological visions of the City of Man is not a rejection of technology and the city, but the joyful obedience to the one true God, who alone is able to save us from ourselves and the terrible gods of our creation, and who alone is able to make our works last and give them value – to give them, in other words, those same godlike qualities that we had sought apart from him. ➤

# Editors' Picks

### Birding to Change the World
A Memoir

*By Trish O'Kane
(Ecco Press, 368 pages)*

In a recent email to me about climate grief, theologian Hannah Malcolm offered a personal note about the approach she takes with her young daughter: 'I have begun to face the responsibility of teaching her to *name* and relatedly *love* the world around her', Malcolm wrote. 'As the world breaks down around us, we will *all*, one way or another, have to make home again when home either becomes unrecognisable or inhospitable. The practice of naming – with all its complexity and tension – offers one way in.'

When I received that email, I was reading *Birding to Change the World* by Trish O'Kane. Malcolm's words reflect the book's central claim: learning to name can foster love that can save a place. The loss of home came for O'Kane just after she had found it. After working as an investigative journalist, O'Kane and her husband moved into their New Orleans neighbourhood in July of 2005, a short walk from Lake Pontchartrain. A month later their house was underwater as Hurricane Katrina overwhelmed the levees.

The upheaval of Katrina drove O'Kane to seek solace in the wild world around her. And in her observation of the natural world, she began to ask new questions. 'Before Katrina, my question had always been, *How can I make the world a better place? What can I do?*' she writes. After Katrina, she took a more contemplative stance: 'I could feel my question changing from *What should I do?* to *How should I be?*' She developed a yearning to understand the life systems of the earth, a desire that drove her to an improbable midcareer move to Madison, Wisconsin, where she began a doctoral programme in environmental studies.

On a freezing day, doing her ornithology homework in Warner Park near her house, O'Kane saw a single, red northern cardinal singing at the top of a tree – a bird she could name. 'It was just me, the cardinal and that cold, bright moment.' Her heart filled with love for that bird and that place. And it changed everything.

'Love will save this place,' writes the activist Naomi Klein of our planet. And this was true for O'Kane, whose love for the creatures of Warner Park drove her to organise a community effort to save it after the city wanted to 'improve' the park by paving over and clearing out its most wild places. The heart of the book follows O'Kane's efforts to build Wild Warner, an organisation working to preserve the wildlife she has come to love so deeply.

In the face of ecosystems in a spiral of collapse, it is easy to become overwhelmed with the question, 'What am I to do?' But O'Kane's rich and grounded memoir shows us that in asking, 'How should I be?' we open ourselves to falling in love. And love can change the world.

—*Ragan Sutterfield, author,*
Wendell Berry and the Given Life

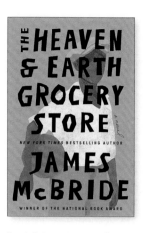

## The Heaven & Earth Grocery Store
### A Novel

*By James McBride
(Riverhead Books,
400 pages)*

1930s Pottstown, Pennsylvania, is home to a multicultural population: Jewish immigrants from Europe, Black families of the Great Migration and working-class Italians living together uptown, wary of the WASPs downtown. A kosher grocery store serves as a meeting place, newsstand and safe haven until an act of violence changes everything.

James McBride's new novel includes multiple disabled characters, a rarity in literary fiction. As a deaf reader, I was drawn towards the portrayals of disability and there is a lot to commend. Chona, though 'crippled', is intelligent, generous and committed to her community. Chick Webb is a talented musician despite being a 'hunchback'. The town doctor and proud Klansman Doc Roberts's role as town villain adds nuance too; he is educated, a bigot, a doctor, an abuser and his disability does not negate his humanness, even when that humanity is evil.

Then there's Dodo, a Black, deaf 12-year-old who lost his hearing in an accident three years before the narrative begins. His portrayal is a bit more uneven. Some of Dodo's representation feels authentic – the way he notices vibration and very loud sounds, how his brain still enjoys music though he no longer hears it the way he used to. In other ways, he exhibits the standard tropes of deaf characters written by hearing people; though he's only been deaf for a couple of years and has had no training or schooling, he is a magical lipreader.

Dodo's schooling presents the biggest plot hole – he has been sent by his aunt and uncle to live in hiding because the state is looking to institutionalise him, an orphan, at the Pennhurst Asylum. When that plan goes awry, Chicken Hill residents pull together to save him. Except the state-run Pennsylvania Institution for the Deaf, an integrated school, was operating in Philadelphia at the time, and evidence suggests that Black deaf students attended before the Civil War. Thomas Flowers, class of 1895, went on to become the first deaf man to graduate from Howard University.

Should it matter, in fiction, if liberties are taken? Not if they're a matter of pure fiction. But in a book that recreates the scenery of Pottstown and the surrounding Montgomery and Berks counties with painstaking realism and includes the real-life musician Chick Webb, this sequence stands out, leaving the reader to wonder whether this was an oversight or a choice to make Dodo simply more plot vehicle than person.

The times we are closest to Dodo's perspective seem intended mostly to satiate the voyeuristic desire to see Pennhurst. We meet a cast of faeces-throwing, paedophilic, mentally unwell men, as well as another teenage boy whom Dodo befriends, even as he bestows upon him the unfortunate nickname 'Monkeypants' and spends the better part of a chapter perseverating on the way in which his limbs are 'tangled up like pretzels' from spastic cerebral palsy.

Such observations are unlikely to deter the average reader from what is ultimately a compelling novel, including some bright spots in representation of the multiply marginalised. But if the point of the novel is to demonstrate human ties that bind, characters must be fully human to be part of the tapestry. Perhaps this, too, is another kind of realism, if not in the facts of the narrative, then embedded into the book's marrow – as is so often the case when it comes to disability inclusion and representation, the results are an uneven weave.

—*Sara Nović, author,*
*True Biz (2022), Girl at War (2015)*

## All Things Are Too Small
### Essays in Praise of Excess

*By Becca Rothfeld
(Metropolitan Books,
304 pages)*

There's something
deeply human about
the biblical story of
King David offering
to build a palace for God. Like David, we all desire
to grasp the infinite, to build something so grand
that it reflects some of what heaven must look like.
But David's suggestion is met with a half-rebuke
from the Lord: 'Would you build a house for me
to dwell in?' Eventually, God agrees that David's
son Solomon can build the Temple. But the Lord's
point to David still stands: outside of heaven, all
things are too small to hold the infinite power,
love and knowledge of the Lord.

*All Things Are Too Small*, an essay collection
by *Washington Post* book critic Becca Rothfeld,
offers an impressive critique of the contemporary
tendency towards smallness and reductionism.
Increasingly, Western culture prizes utility and
efficiency over beauty and magnificence. 'We are
inundated with exhortations to smallness: short
sentences stitched into short books, professional
declutterers who tell us to trash our possessions,
meditation practices that promise to clear
the mind of thought and other detritus.' This
tendency often places value on things not for their
aesthetic or moral qualities, but for their efficiency
and cost-effectiveness. Of course, there can be
beauty in smallness – the tiny flower or the intri-
cate carving. But smallness, driven only by utility,
leads to mediocrity, not beauty.

Rothfeld's writing is honest, beautiful and
at times quite funny. It spans a wide range of
subjects from her favourite novels, essays and
murder mysteries to mental health and Simone
Weil, but her plea for a more expansive view of the
world, her desire for the infinite capacity of divine
love, threads its way through it all. 'Does heaven
exist?' she asks. 'I don't know… [but] heaven is real
in one important sense: what we demand from
it reveals so much about… what we desire.' If the
heaven we imagine is a place of love and self-sac-
rifice rather than greed and hatred, wouldn't we
try to live out those values now? Why wouldn't
we fight for beauty, truth and goodness? After all,
'there is nothing admirable in labouring to love a
world as unlike heaven as possible'.

Many of the mystics, medieval and modern,
whom Rothfeld praises throughout her essays
embraced poverty and fasted to the point of
starvation in their desire to be closer to heaven
and to Christ. This asceticism might seem at
odds with Rothfeld's defence of excess and
magnificence, but she argues that their actions
were driven not by a longing for smallness, but
rather by a recognition that earthly goods were
insufficient in satiating their desires. For some, the
only food they ate was the Eucharist – only Christ
himself could fill them.

Rothfeld returns to and lingers on the vastness
of love. 'How long is the conversation of love?'
she asks. 'Nothing less than everlasting would be
enough. Not all loves *do* endure, of course, but it
is essential to love that we believe it will, that we
want it to, because an orientation towards eternity
is part and parcel of what it is to have faith in the
inexhaustibility of another person.' This orientation towards eternity is vital. If our fellow humans
are inexhaustible, how much more vast must God
be? Like the mystics', our desire for the infinite
cannot be quenched by earthly, mortal things. All
things are too small to quench this desire except
for the Lord of heaven and the kingdom over
which he reigns. As Augustine famously wrote,
'You have made us for yourself, and our hearts are
restless until they find their rest in you'.

—*Alan Koppschall, editor,* Plough Quarterly

# A Church
# in a Time
# of War

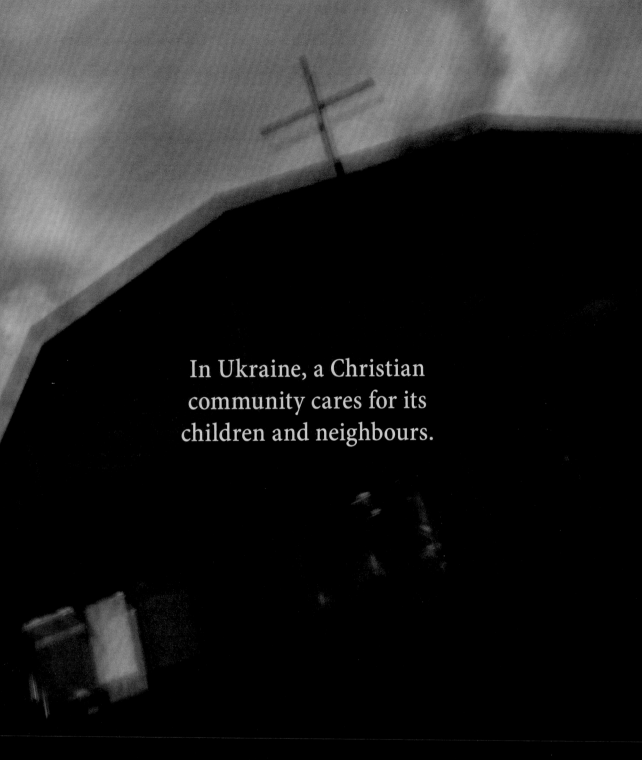

In Ukraine, a Christian community cares for its children and neighbours.

**AN INTERVIEW WITH SASHA RIABYI**

*With photographs by Danny Burrows*

*In March 2024, photographer Danny Burrows travelled to Kherson Oblast in Ukraine with the NGO Novi. His assignment was to document their work 'restoring child-hoods disrupted by war' in villages along the frontlines. Here, he met Sasha Riabyi, a Novi representative and a pastor of a thriving church community in Khotiv, on the outskirts of Kyiv.*

**Danny Burrows: While documenting your work with Novi in Kherson, Ukraine, I found it remarkable how central the church has become to Ukrainian communities on the frontlines of this war. What is it like to live and work in Ukraine at this time as a Christian?**

**Sasha Riabyi:** We are normal people. There's nothing special about us. We're just lucky to live in Ukraine during this time.

**Lucky?**

Yes, lucky. Many have told me this. We helped each other. We started doing something for others, for all Ukrainians. And that, as a nation, was a very important change. You can't make it artificially. I say we're lucky because it's amazing to experience this.

**It's an awakening for individuals and as a nation?**

Yes, we woke up. It's a very good description of what happened with this war. For over two years now, we've been living knowing that tomorrow or today, life can change at any moment.

In one second, things can start falling apart. You go to McDonald's, to church, even where people are getting free water and food parcels, anywhere where there's a crowd of people. In 20 minutes, a missile could hit that same spot.

**Can you describe the first day of the invasion?**

On the morning of 24 February 2022, we were awakened by explosions, the sound of missiles hitting the ground. There was heavy bombing in Kyiv. That's where we're located.

My wife and I woke up, and our first thought was to protect our children. I went to their room and said, 'OK, wake up. Let's go'. My oldest boy asked, 'What's happening?' And I told him, 'Putin is making some extra noise here, and that's why we need to go to a safe place'. Because how do you tell a child that a war has started?

Our own families are the most important thing for us each to take care of, but we also live in community – our church is our community. And we all agreed before the war that if something happened, we would all come together at the church to see each other, to plan something.

*Danny Burrows* is a professional photographer and journalist whose work has appeared in the Times, the Independent, *the* Guardian, *and* Another Life Is Possible *(Plough, 2020).*
*Sasha Riabyi* is a pastor, husband, and a father of two. He lives with his family in Khotiv, a village just outside Kyiv, in Ukraine.

*Previous spread:* The church in Khotiv, at which Sasha Riabyi is pastor.

At a retreat organised by churches and the NGO Novi, Sasha Riabyi entertains children of active servicemembers and soldiers who have died at the front.

A kindergarten classroom battered by Russian shelling on the frontlines. The Russians have targeted civil buildings such as schools, churches, theatres, and halls. There are few of these buildings that are not damaged.

We split our church into two groups. The younger group, around fifteen people, left in one direction all together. We had seven cars. Another group of older people did not want to leave. 'We are too old for travel,' they said. 'If the occupation comes, we will just stay here. Here, we have our home.'

We went to Western Ukraine, and we stayed there for three months. But when the Ukrainians pushed back the Russians, we started thinking that we needed to return home to try to take care of the people in our community. We came back and did what we could to help. It was very necessary. There was no food on the supermarket shelves, and prices were very high.

**Your whole ministry really changed. It became a ministry to a wider community.**

It totally changed our church. It woke us up. It woke me up as a father, a citizen. It woke me up as a pastor.

We opened the doors of the church. It's not only for services and prayer meetings now, but people come for projects, to play games, to interact with each other, to have a meal together. They can pray together as well. It's now open six days a week, like a community centre.

**Your ministry also built a new building during the war.**

Our church was around 25 people, a very small community really. And then suddenly we grew. When you have 70 children and 300 adults coming regularly, what can you do? We tried to think what the biggest need in our community was and decided that the children were the most important. We needed a space for the children. So,

we asked God, 'If this is from you, open the doors'.

About two weeks after we started praying, one of my friends called me from the United States. He asked, 'Are you doing something to help these refugee people in your area? Because knowing you, you won't sit with your hands down'. He gave us around two-thirds of what we needed. In several months, we already had the walls, windows, and roof up. Other people noticed what we were doing and started helping us. Then, we added a kitchen to feed people.

**What's the weekly routine at your ministry now?**

On Mondays, internally displaced people come. On Tuesdays and Wednesdays, local people come. We sit at small tables, and we share food and conversation. You cannot imagine how much grief and sorrow is in their hearts. People start talking and can't stop. It's given us a chance to meet many people, to know their lives, to know them. Many have become close friends.

We always have questions prepared. 'What was the worst thing that happened to you last month?' 'What was the best thing that happened to you last month?' We speak about anger and what to do with your anger, how to fight it. We discuss how to put it into practice.

Recently, one lady told me, 'Sasha, I want you to know that these discussions are helping me to fight this anger'. All people in Ukraine are angry. Even if we're smiling, deep down, we're angry, because war is evil.

On Thursdays, a ladies' group comes for a meeting. Every second week we have a breakfast for young mothers, and they discuss how to raise children. Raising children is hard everywhere, but

The remains of a Russian missile that killed an eight-year-old girl who was sheltering under the kitchen table of her home in Kherson.

Children carry the outline of a child drawn during a 'Helping Hands' session
run by Novi in a bomb shelter under a school in Kherson.

in Ukraine the war makes it harder.

Friday is a day off (we need a day off). But then on Saturday we invite children from the community. They can play games, draw pictures, do homework, have a meal – a sandwich, a pizza, maybe sometimes soup or hot tea.

Also, we run a Young Peacebuilders Club. The idea is very simple. A group of teenagers get together to find solutions to the broken world around them that they can put into action. Give teenagers a little bit of energy in the right direction, and they go far.

And Sunday, of course, it's a church service.

**Can you describe the humanitarian work that you are helping with?**

When the war began, getting humanitarian aid was very easy. But food parcels do not solve every issue. The trauma of the children, for example – food cannot help with that. Novi said, 'We want to help with the children because we know that trauma lives on and affects the future of the nation'. Their experience, their love and care, was greatly appreciated. They started bringing in their 'Life Kits', developed by a Norwegian psychologist. The kit is a backpack filled with toys and an instruction book. These kits are designed to help children feel more confident and safe. They guide children in attention-based, physical, and interactive play, helping to relieve stress and build skills through positive activities.

Now we're helping Novi all over Ukraine, distributing these simple toys and setting up programmes such as the Young Peacebuilders Club. There are more difficult areas than ours. In Central and Eastern Ukraine, children see bombings; they see missiles. It only takes one explosion to scare a child. Even if they only hear it, it stays in their hearts and minds. It's scary.

**When the war is over, there will be a lot of trauma to deal with, both within the nation and within individuals. How do you see your ministry continuing? How will it change?**

Sometimes we dare to think about when the war will end. We understand that many traumas will live on. They show up in communities already. I can tell you for sure that we're not ready for that. I don't know if any country would be ready for that. But at least there should be a place in every community in Ukraine, in every town and village, where men and women who lost their parents, spouses, children can come and at least talk to each other. I don't know if we can help them when they come, but we need to give them a place where they can be listened to.

We see children who are very traumatised. We ask about what brings them joy and they say, 'A dead Russian soldier'. We see a boy smiling because he drew a picture of himself holding a knife with blood and he says, 'Oh, this is me, and I killed a Russian, and that's why I'm feeling good'.

Children see the warfare; they see the hatred. If we don't take care of them right now, can you imagine Ukraine in 10, 15 years with these children who saw all this?

It's only through community, through small groups of people, that change comes into our lives. Through our neighbours, the person I trust, the people I know – that's how change is going to come. ➴

Children of active servicemen and soldiers who have died at the front attend a retreat in the Carpathian Mountains organised by Novi.

## Fingered Forgiveness

I.
Once your mother, always your mother
*Sine moj*, I want to swaddle you
in the hand shorn slippers of my youth.
All that I have to give, to be
in tattered, sun-kissed pockets
crammed with sugar cubes
not soaked by a sour separation.

From wind tipped grasses of a hill country
teeming with pickled cabbage and drooping plums
to slouching shadows in towering high-rises
exhausted from the daily pursuit
of circling roundabouts without an exit,
I have come and find myself lost

in between the *lj, đ, č, š,* and *ž.*
The shadow of my former self swarms
about the vast void seeking these sounds
with their tongue twisting form.
A child of wobbly words in your world
I have become.  Its forgetfulness in you
I refuse *iz inata* to mourn.

II.
In a cellar hidden from falling grenades
I found myself.  Barred from flicking my lucky marble,
sea green nicked by tick of the clock,
I giggled at *moja* Dalida and picked up another
to play. "Clink!"  One after the other they rolled
in the candlelit dust.  She stomped off forever
forging her own *roša* into the crumbling cement floor.

III.
The mother of your mother
Her voice, in the tunnels where my thoughts wade
through muddled memories, faintly still waxes and wanes:

    *Pogledaj im kroz prste, sine moj.*

LAURA R ECKMAN

**NOTES:**

*sine moj*: Literally 'my son', but in Balkan culture it can also imply 'my daughter'.

*lj, đ, č, š,* **and** *ž*: letters of the Bosnian-Croatian-Serbian alphabet, the *lj* is like the 'lli' in million, the *đ* is the 'j' in jump, *č* is 'ch', *š* is 'sh' and *ž* sounds like 's' in the word 'measure'.

*iz inata*: out of spite

*moja*: 'my'; when used with a first name, it can speak of a close female friend.

*roša*: a hole found or created for the Balkan version of marbles

*Pogledaj im kroz prste*: an idiom which literally means, 'Look at others through the gaps of your fingers'. Figuratively, it means to overlook another's mistakes.  ⤝

# Towards a Gift Economy

*Some goods and services have value beyond their market price.*

**SIMON OLIVER**

I N MODERN DEMOCRATIC SOCIETIES, our lives are dominated by markets. We buy and sell goods and services, take out loans and mortgages from banks, make investments, contribute to pension schemes and pay taxes to the government in exchange for public services and the protection of our rights and liberties. We sell our labour on the labour market. We have credit ratings that signify the risk we pose in the market for credit. Universities market themselves based on the projected earnings of their graduates so that a degree becomes nothing more than a shrewd financial investment. Projected lifetime earnings indicate our value in terms of what we will earn and what we will pay in taxes. The market for healthcare places monetary value on the quality and length of life. The ledger of modern life is expressed in terms of trade, credit, debt and account.

In his 2012 book *What Money Can't Buy: The Moral Limits of Markets*, the American political philosopher Michael J Sandel points out that,

Pam Ingalls, *Julie's Table*, oil on board, 2018.

United States, some Western couples outsource pregnancy to surrogate mothers in India, where the cost is a fraction of the going American rate. Lobbyists wishing to attend a congressional hearing on Capitol Hill might not want to queue overnight in the rain, so a market has opened up allowing them to pay others, predominantly the homeless, around 20 dollars per hour to queue on their behalf. It is possible to sell parts of one's body as advertising space. Some pupils in underachieving schools are paid to read books.

The pervasiveness of markets and our evolution from having a market economy to being a market society are not inevitable developments of human culture. They require a particular imaginative framework and worldview. One characteristic of the modern world, beginning around the 17th century, is the worldview that separates nature and culture. Nature becomes a domain of objects with no intrinsic value or orientation except trade and consumption within the cultural domain of market societies. If nature has no intrinsic purpose or value, no origin, end or meaning, its value lies in what we can get for it – in trade and consumption. The idea, therefore, that humanity, having risen above nature, can manipulate and exploit nature to its own ends through technology provides a framework for the rise of capitalist market economies and industrialisation. Is there an alternative imaginative framework, a different way of understanding humanity's relationship to nature, a more fundamental economy?

Despite how pervasive markets have become, the relationships and social units that we most value and prioritise, particularly friendships and the family, do not primarily involve the market, money or contracts. They are expressed through a different economy, a gift economy. A friendship, for example, is a relationship based on sharing life

while trade and currency are primitive aspects of human society, the reach of markets accelerated rapidly in the late twentieth century, particularly under the influence of the laissez-faire economics of Margaret Thatcher and Ronald Reagan. We have moved from *having* a market economy to *being* a market society in which almost anything is tradable. Sandel cites the reach of markets. In certain prisons in the United States, for example, it is possible for nonviolent prisoners to purchase a prison cell upgrade for around ninety dollars a night. They receive a clean and quiet prison cell, undisturbed by nonpaying prisoners. While commercial surrogacy is illegal in the United Kingdom and highly expensive in the

*Simon Oliver is an Anglican priest, a theologian, and the Van Mildert Professor of Divinity at the University of Durham. He lives in the United Kingdom.*

and the gifts of time and attention. The family is a web of relationships established on gift exchange. A father does not charge his children for reading to them or cooking their dinner, for both are gifts. A wife does not charge her husband for caring for him when ill, for care is a gift. Indeed, we talk of 'care*givers*'. When children are taught to thank those who protect and provide for them, their thankfulness is an acknowledgement that what they have received is the gift of loving care; their gratitude is a return gift. We laud those who donate (rather than sell) their time and skill to their communities, for example in running a Scout troop or volunteering in a charity shop. We prize philanthropy. The church is a body

---

## The gift is not a mere object or commodity; it bears meaning and mediates a relationship.

---

overwhelmingly constituted by the gift of people's time and talents. Intimate relationships such as marriage involve the gifts of love and attention that are never reducible to trade, debt and account. One of the many reasons why the breakdown of intimate relationships is so painful is that a bond previously expressed in terms of shared goods and the exchange of gifts is often broken down into monetised assets that must be divided and debts that must be paid.

Contemporary discussions of the gift economy have roots in the work of the French anthropologist Marcel Mauss and his book *The Gift* (1923). Mauss analyses fieldwork concerning the practice of gift-giving and gift exchange in archaic indigenous societies, particularly in Polynesia, Melanesia and the American Northwest. These societies, where traditional practices of gift-giving and exchange were still relatively untouched by modern capitalism, led Mauss to two important conclusions.

First, a gift conveys something of the giver to the recipient. Take a simple example: At nursery, a child paints a picture for her mother and gives it to her when they arrive home at the end of the day. The gift – the picture – expresses the child's imagination, skill and view of the world to her mother. The gift is not a mere object or commodity; it bears meaning and mediates a relationship. If a friend gives me the gift of time and attention in listening to my hopes and fears about the future, she is conveying her sensitivity, thoughtfulness and concern – important aspects of her character, perhaps fashioned over many years. The time and attention given by my friend express and confirm the friendship. As Mauss writes, 'It follows that to make a gift of something to someone is to make a present of some part of oneself'.

According to Mauss, the giving of a gift can be contrasted with the transfer of private property in a market economy. When one sells a car, ownership rights are transferred from the seller to the buyer. The seller is alienated from the object sold and has no further claim upon it. The relationship between seller and buyer lasts as long as a contract of sale remains unfulfilled; once the transaction is complete and obligations satisfied, they go their separate ways. Gift-giving is quite different. Because the gift carries with it something of the giver, it continues to be imbued with power and significance. The giver is, in a sense, present with the gift. For example, the girl's picture will always be, in some sense, the girl's picture that she gave to her mother. According to Mauss, this means that gifts establish lasting social bonds between donor and recipient.

Because gifts bear meaning, their value to donor and recipient cannot be expressed straightforwardly in monetary terms. Take the following simple example. On my desk, I have two objects: a computer I bought for my work and an engraved fountain pen given to me by my brother when I took up a new job. The computer is extremely useful. It was expensive and, because it still functions quite well, I could sell it for a reasonable amount of money. It has a monetary value and I know how much it would cost to replace. It would be inconvenient if the computer were lost or broken, but I could replace it quite easily, assuming I had the necessary money. The computer has a function and a monetary value as a commodity, but it has no meaning, because it does not mediate a relationship. The pen, on the other hand, is a gift. It conveys something of my brother to me, namely our shared liking of traditional things and my brother's recognition of me as a writer and academic. The pen makes a claim on me – I look after it and enjoy it and I am reminded of my brother and my vocation whenever I use it. If I lost the pen, it could not be replaced. I could buy a similar pen, of course, perhaps an identical pen, but it would not be the pen given to me by my brother. If I were to attempt to sell the pen, which would feel like a rejection of my brother and his gift, it would probably fetch a fraction of its price when new. The pen is not a tradable commodity. Whereas the computer's value lies in its utility, the pen's value lies in its utility *and* its meaning. Gifts really are magical; they mediate the giver to the recipient and hold a power over us.

This leads to Mauss's second conclusion concerning gift-giving: because true gifts establish and mediate social bonds, they are reciprocal. A gift, far from being unilateral, awaits and expects a response in the form of a return gift. So, it is not simply gift-giving that is important in the societies studied by Mauss, but gift exchange.

When I give the gift of a meal to my friends, for example, I hope for a response in the form of a return gift. That might take the form of a bottle of wine donated at the start of the evening, a thank-you card the following day or an invitation to dinner in a month's time. This exchange of gifts establishes a relationship of charitable friendship and invites a pattern of gift exchange in which donors and recipients continually convey something of themselves in the gifts they offer. This is one way that intimate relationships and family life are formed and expressed – members of the family continually offer each other favours (doing the laundry, cooking a meal, arranging a birthday party, organising a picnic, buying Christmas presents) and children are initiated into this round of gift exchange or 'economy of favours' as they take their place within the daily life of the family.

When we read the Christian and Hebrew scriptures, 'gift' is a concept that hides in plain view. We might not notice its significance, but once our attention is drawn to the prominence of gift-giving and receiving in the story of Israel and

Pam Ingalls, *Win Win*, oil on board, 2015.

the church, as the theologian John Milbank points out, we see gift as an all-encompassing theological category. Christ is God's gift of himself to creation (John 3:16). The Holy Spirit is known as 'the given' (*donum*) amongst theologians of the early church (reflecting, for example, Isa 11:2–3 and John 20:22).

# Life is not a matter of trade, possession, or right, but of gift.

Importantly, according to Saint Paul, the church is the community of the gifted (Acts 2:1–13; 1 Cor 12; Eph 4:11–13) and grace is God's gratuitous gift for our salvation (Eph 2:8).

Most fundamentally, creation is understood as gift, for all things come from God (James 1:17–19). By the early second century, through attentive reading of scripture and careful philosophical reflection, Christian theologians (in common with Jewish and in due course Islamic, thinkers) brought a new clarity to our understanding of creation: God creates all things *ex nihilo* – out of nothing. If we are to talk intelligibly of God and his unique creative act, we must not think of the universe as existing in endless time. Nor must we think of God fashioning the cosmos from a primordial soup. God is the source of everything that is not God, including space, time and matter. Everything has an ultimate beginning in God's creative act. Creation is therefore *the gift of existence*. Reflecting themes in the anthropology of Marcel Mauss, we can say that creation bears something of the giver, God, to the recipient, creatures. It is an expression of divine beauty and goodness. Creation also invites a response in the form of a reciprocal gift of thanksgiving. This exchange forms a relationship of covenant between God and creation.

Within the order of creation, the gift of life is particularly important: the life that God gives in creation and breathes into Adam (Gen 1:11–12, 20–25; 2:7); the gift of food to sustain life (Gen 1:29); the life to which God continually recalls the people of Israel (Deut 30:19; Ezek 18:32); the life that God gives in Christ, who is the resurrection and the life, the way and the truth, and who gives his flesh for the life of the world (John 11:25; 14:6; 6:51). Christ came that we may have life and have it abundantly (John 10:10) and through Christ God calls us to eternal life (Rom 6:23). The Christian difference is this: To receive our lives as a gift rather than as the outcome of a blind evolutionary process or as a mere possession. We come to know that our life as gift conveys something of God, the donor, to the living; it invites a response. Despite sin's refusal of the gift of life, it is given again in the waters of baptism. Life is not a matter of trade, possession or right, but of gift. My life is *my* life, given by God.

The Christian imagination therefore knows little of the distinction between nature and culture, for both are encompassed within the more fundamental concept of creation as gift. Nature is a gift of creation within which the human person receives her life as gift. Human cultural life and creativity, our polity and economy, depend entirely upon and are enfolded by, nature. Nature provides the gift of food to sustain the gift of life. Most importantly, if nature and culture are enfolded within creation, both have an intrinsic meaning and purpose as gifts of God and expressions of divine goodness. Trade may be necessary and markets inevitable, but there can be no ultimate *commodification* or *reification*, because market economies are always reliant upon the more fundamental economy of creation as gift. This theological imagination can provide a very different framework for reintegrating the natural and the cultural and renewing a sense of humanity's place and vocation within a *created* order and a divine economy of gift in which meaning, purpose value are restored. ➤

# Gerhard Lohfink: Champion of Community

*We don't follow Jesus alone.*

**TIMOTHY J KEIDERLING**

GERHARD LOHFINK, CATHOLIC PRIEST, theologian, prolific author, *Plough* contributor and friend, died on 2 April 2024.

After receiving a doctorate in theology from the University of Würzberg in 1971, Lohfink taught at the University of Tübingen. In 1987 he decided to leave his professorship and move to join a group of like-minded Catholics living in community in Bad Tölz, Germany.

Several years after he joined the community, he began to write theological works again. In all, he published 22 books. He was at work on an autobiographical manuscript, 'Why I Believe in God', less than a month before his death from a severe illness.

I will always be grateful to Lohfink for his friendship. In 2019 I spent five months as a guest of the small house community of which he was a member. Over our daily dinners we would talk about theology and life. We'd then wash the dishes together and continue the conversation. Sometimes he would invite me to join him on his long walks or to visit his favourite churches; other times, we'd sit and pore over texts in his office. When I last spoke with him, in January 2024, his health had deteriorated significantly, but he still spoke with the same quiet charisma and fire about his latest book project.

Lohfink's greatest insight – the one that determined the trajectory of his life – was that following Jesus can only be done with others. This comes to expression in many of his works. In a piece he contributed to the *Plough* anthology *Called to Community: The Life Jesus Wants for His People*, he writes:

Christian faith, just like Jewish faith, subjects all of life to the promise and claim of God. Its nature is such that it interpenetrates all aspects of the lives of believers and gives them a new form. Of itself it demands that social relationships must change and that the material of the world must be molded. Faith desires to incorporate all things so that a 'new creation' can come to be.

At the same time faith tends towards a more and more intensive communion among believers, for only in the community, the place of this communion, only in the place of salvation given by God can the material of the world really be molded and social relationships really transformed. It would therefore be essential to Christian faith that individual believers should not live alongside one another in isolation but should be joined into a single body. It would be essential that they weave together all their gifts and opportunities, that in their gatherings they judge their entire lives in light of the coming of the reign of God and allow themselves to be gifted with the unanimity of agape. Then the community would become the place where the messianic signs that are promised to the people of God could shine forth and become effective. ⇜

From Gerhard Lohfink, *Does God Need the Church?* (Liturgical Press, 1999). Used by permission.

---

*Timothy J Keiderling lives with his wife and two daughters at Woodcrest, a Bruderhof in upstate New York.*

# When a Bruderhof Is Born

*What's it like to be a young person in a young community?*

**MAUREEN SWINGER**

OCCASIONALLY, FRIENDS HAVE ASKED how new Bruderhof communities begin. The short (and possibly unsatisfying) answer is that each one has its own story. I can only tell the story of the one I got to witness firsthand.

An hour and a half's drive north of New York City, in Chester, New York, lies the Bellvale campus. The property became a Bruderhof in 2001, but it came with a history. Below the ridge of Bellvale Mountain, a sprawling main building with a graceful A-frame chapel and three identical brick cottages run in a line across the valley, fronted by a 16-acre lake.

It started out as a welcoming place in the early 1960s, when Pius XII Youth and Family Services opened its doors to wards of the state, young men who had no other safe home. Run by the Holy Cross brothers, this boarding school provided education, stability and care to hundreds of young men for the next 30 years. We know this firsthand because in our first ten years of residency there, we hosted reunions for Pius XII alumni and

*Maureen Swinger is a senior editor at* Plough *and lives at the Fox Hill Bruderhof in Walden, New York.*

Pius XII Campus, now Bellvale, in August 2002 – one year after it became a Bruderhof. All photography by Tim Clement.

Neighbours complained of vandalism and threats to safety; police were frequently called in to address fights or runaways or property damage. With so many troubled kids, a shortage of trained staff and no funding to expand resources and repair the campus, Pius XII managers abruptly declared the school defunct in May 2000.

The campus sat empty for a year; the only activity that Bruderhof scouts encountered was generated by rats still hanging around the dumpsters out back, and feral cats prowling about on account of the rats.

The scouts were looking for a place that would not have to be built from the ground up – preferably a former school or monastery that could take about a hundred people within a few months. The oldest US Bruderhof, Woodcrest, in Rifton, New York, was supporting more people than it ever had, and community members were considering the idea of moving the Community Playthings administrative offices and all connected families and singles to a new location. (Community Playthings is a Bruderhof company that supplies wooden furniture and toys for childcare centres.)

This proposal was met with enthusiasm by the families involved. When the first photos of our future home were posted, catching the outlines of the buildings, the mountains and the lake, we all agreed it looked beautiful from the road. But the scout team also reported its state of dereliction. To my mind and to many others', this increased the appeal. How often do settled office workers in a well-established community get to try their hands at pioneering?

In July 2001, the first six people arrived, clearing and cleaning the rooms in the furthest cottage, getting one kitchen and two dorm-style bathrooms up and running, removing the worst piles of detritus and turning hayfields back into something like lawns.

My office – tech support and helpdesk – moved in October. With one family and two singles, we brought the total population up to 20. After

they came back in droves, transforming on arrival from dignified lawyers, architects,and chiropractors to gregarious teens who tore around campus finding their old rooms and favourite fishing spots. Under a willow on the far side of the lake, they stopped to salute the memorial stone of the first  of them to fall in Vietnam – Eugene Kirkland, 6 May 1968.

These men told of the love and high expectations showed them by the brothers, of the camaraderie they found. They were in touch with each other still and acted like family. They declared the property as beautiful as it had ever been and decried the fact that they had only been allowed to gaze at it from the main road for the last decade.

Why couldn't they enter? By the early 80s, just as some of the most dedicated Holy Cross brothers were aging out, the New York courts began referring kids with criminal records to Pius XII. With very few brothers left to run the place and increasingly violent teens being shipped from juvenile centres each week, Pius XII had to hire staff who doubled as security guards.

scrubbing down our own office in the main building, we plugged in the computers and phones. We two single ladies plopped mattresses down in a room in the middle cottage. The cottages were all the same, sporting three desks built into cubbies in the wall, three metal stools, each chained to a desk and three closets, identical except for the gang graffiti scrawled on the doors. (My roommate was apparently assigned to the Kings and I to the Bloods.) The windows were shatterproof and also viewproof Plexiglas, with steel grates on the outside.

We pried off the grates with a crowbar, but the windows were screwed shut. I tracked down an electric screwdriver and freed up the window frame, only to have the top half slide down at guillotine speed and pulverise my knuckles before I could react. My bid for sympathy was met with laughter from older pioneers – it seems this accident had happened to multiple other seekers of fresh air and had become a rite of passage.

Keeping an office headquarters humming so that distant factories didn't experience a lag in service, while also tackling various corners of our wonderful, sprawling, messy new home, proved a rousing exercise in balance. The flair and fun of winging it and sometimes blundering in the attempt, and apologising and taking another run at it, was exhilarating. Over the next weeks, more young singles arrived, sent from other communities to help conquer the place.

I had been an introvert in my last home; the community was large and the youth group comprised about 50 people who had known each other most of their lives. I was also trying to let go of a wonderful friendship. Maybe we both had hoped for more, but it ended in distance and silence. I came to Bellvale ready to put my head down and work, but probably skip the socialising.

That's not how it ended up. This youth group, soon numbering around 20, was thrown together so haphazardly into common purpose as we tackled dysfunction and grunge that it was impossible to be a wallflower. We got to know each other while running late-night cleaning projects, chipping graffiti off the walls, hurling rocks to knock pairs of old high-tops off the telephone lines, and dismantling the 12-foot-high perimeter fence.

Between work projects, we found opportunities for downtime and here the old campus was in our favour. Each of the three residential cottages had a big, central foyer, complete with fireplace and a circle of sofas. In the evenings, all you had to do was stroll the path fronting the cottages and pick your entertainment via the picture windows. In one foyer, an earnest political discussion. The next, a boisterous sing-along session. The last, an intense poker game. Folks mixed easily between houses and lines between age groups or families versus singles, were happily blurred.

There was more laughter in that first year than in most of our lives before or since. Bizarre situations arose daily. One morning someone opened an electrical panel in the main house basement and

Investigating a faulty heater in the main building.

found a mummified cat. He put it in a shipping box and addressed it to his uncle, a prankster himself, who reacted not at all when opening the box, but made the most of others' reactions as he showed his 'gift' around at snack break.

Someone else put his head through a newly replaced window. It was a normal, clear glass window, but we were all so used to the clouded, scratched reform-school specials that he assumed his ability to see through it meant it was open. After getting stitched up, he reported his blunder at communal lunch. We couldn't help laughing along with him.

An intrepid sewer-plant investigator found a disgusting pair of neon-orange plastic overalls. From then on, by unspoken agreement, those trousers appeared in every Saturday-night skit, part of our weekly chance to cut loose and laugh over recent efforts and setbacks.

Our evening meetings were often lively discussions as to what was the right way forward or the most important project out of the many clamouring for attention. There were disagreements and misunderstandings. But we gained new appreciation for our century-old community rule for direct speaking, with no gossip or undermining of others.

No one could forget Bellvale's first Christmas. Many of the trappings and traditions accessible in a more settled community were simply not there. The teens joined the youth group in a simple interpretation of the Christmas story, as Joseph and Mary knocked at the seven different doors opening into our dining hall and were turned away by various modern-day innkeepers, from an argumentative church auxiliary meeting to an elegant hotel concierge to a representation of our own community, too busy trying to fix and clean and build to welcome in the strangers. The young couple found refuge in a room behind a rowdy bar, as each of the formerly busy or proud innkeepers felt convicted of their blindness and turned up belatedly to worship the baby King. By the end, our whole community had gathered before the child.

IT TOOK ALMOST A YEAR till the Community Playthings office was fully staffed, in addition to the teachers, chefs, medical personnel and everyone else needed to keep a Bruderhof running. We now had a functional dining hall, communal laundry, school classrooms and a garage. A feisty grandma by the name of Sibyl joined us, declaring that her purpose in moving there was to give us a chance to start a cemetery. (She proceeded to live – gleefully – for another 12 years.) Now we had all generations represented, from grandparents down to one new baby.

Traditionally, singles on a Bruderhof join up with a particular family for weekends or holidays. I was very lucky to land with Milton and Sandy, who had raised a large family and been blessed with many grandchildren, most of whom resided elsewhere. The one daughter at home, Lisa, who had Down syndrome, was in her late 30s at the

time. They treated me like another daughter and over the course of five years, Lisa and I certainly became sisters, tag-teaming on art projects or baking endeavours, and starting out each workday with an hour in the communal laundry, where I ran the industrial machines while she folded cloth diapers and kitchen aprons, both of us singing loudly over the churning of the washers.

One summer afternoon we were sauntering along the path together, watching a soccer game on the field below. We stopped to identify a goalie who was making saves and blocking shots with death-defying jumps and dives. We were highly entertained. Lisa informed me that this was Jason Swinger, who had just moved to Bellvale from the Fox Hill community. I had no inkling that the goalie and I would be getting married in four years.

Singles shared breakfast twice a week; at first the guys were chefs on Tuesday and the girls on Friday. It developed into a serious culinary competition, with people waking earlier and earlier, borrowing waffle irons from various families and snaking extension leads to the tables from every socket in the dining hall or hosting an omelet bar with all the trimmings. Unsure where this would end or if our community budget could sustain it, we eventually called a truce and agreed to mix it up. Whoever wanted to cook would do

so, while the others showed up to drink coffee and scribble silly messages on the menu board. We started hauling our breakfasts farther afield, like the winding boardwalk at the end of the lake, with everyone sitting cross-legged and passing the pancakes along the line. Once there was cake and coffee on top of the mountain.

Jason and I discovered that we both loved jam sessions and often instigated fireside sing-alongs that boomeranged from rock to country to folk and back again till midnight or until some poor exhausted mother stuck her head into the foyer and shut us down. Sometimes we drifted into the acoustically exquisite chapel, a room that makes any singer sound good.

BELLVALE HOSTED A YOUTH CONFERENCE in 2004, with hundreds of young people camping on the fields and discussing visions for the future of the Bruderhof. Many had been wishing for a way to try out the concept of small urban community houses. This inspiration gained footing and soon after, Bruderhof houses did start up in several cities in the northeast. The groundwork for this idea was laid by Justin Peters, an energetic design engineer, landscaper and father, who had been bringing up the concept in meetings for years. He was also the one helping

A winter evening of Scrabble and cards for old and young.

harness the power of the conference youth into a work team to landscape our community's future cemetery, plant a circle of trees inside and build a low stone wall around the grounds.

Just two months later, Justin drove to a nearby event with other Bellvale members, many of them from our youth group, to hear a talk on peace strategies for ending the Iraq War. As he walked towards the entrance, pointing out the striking beauty of a line of columnar oaks, he dropped to the pavement with no warning, suffering a fatal heart attack.

After an urgent phone call, those of us at home quickly assembled to pray for his life, unaware that he was already gone. His wife, Linda, went directly to the hospital, but his daughter was in our gathering when the second phone call came. Our hearts broke together as we realised that a man who had been a part of all our lives, whom we had worked alongside, argued and laughed and sung with, was not coming back. One hour before, he had walked past some of us planting shrubs around the front of the main building, thanked us for the work and offered placement suggestions. How come we didn't know to say goodbye?

That night, as we lined the driveway holding candles while his body was carried home, our close community was united in a new way, through silent shock and pain. The suddenness of his death was shattering. It also opened up a space for us all to remember why we were here, on this small, beautiful corner of the earth: not to build up a perfect, pretty community but to attest to a way of life shown us in the Gospels.

Justin's was the first grave in the cemetery we had built with him. When I walk through that gate now, I don't think of his death or those of the other nine now buried there. I think of the way they believed in the truth of a shared life, for a greater purpose than community itself. Robert Frost captured that duality in the final lines of 'Two Tramps in Mud Time':

But yield who will to their separation,
My object in living is to unite
My avocation and my vocation
As my two eyes make one in sight.
Only where love and need are one,
And the work is play for mortal stakes,
Is the deed ever really done
For Heaven and the future's sakes.

TWO DECADES ON, most of those first hundred residents of Bellvale have moved along to other communities and stages of life. But, just like our Pius XII predecessors on the same campus, when we are reunited all the stories surface again. People start to laugh and relive the work and the craziness and the joy. And someone will always remember Justin. We've reached the point now where our children can tell the stories on our behalf. That's just fine with us, until they in turn get the chance to be part of their own new community. ✈

Justin Peters tells a story to some of the schoolchildren.

# PLOUGH BOOKLIST

**To order: email** *contact@ploughbooks.co.uk,* **or call** *0800 018 0799*

**Subscribers 30% discount**

**Members 50% discount:** Plough Members automatically get new Plough books. Learn more at *plough.com/members.*

## Church History

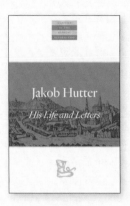

### Jakob Hutter
His Life and Letters
*Edited by Emmy Barth Maendel and Jonathan Seiling*

This scholarly biography and collection of writings by and about an early leader of the Hutterites, a pacifist communal Anabaptist group, sheds light on a persecuted religious minority during the Reformation.

'This volume is of great significance for the study of early Anabaptism. It is essential reading for all who wish to understand – and be challenged by – what Hutter said and did.'
—Ian Randall, Cambridge

Softcover, 388 pages, £22.99 **£16.09 with subscriber discount**

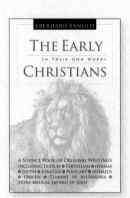

### The Early Christians
In Their Own Words
*Eberhard Arnold*

What did Christianity look like before it became an institution? Find out for yourself with this collection of firsthand accounts of the early church. Includes excerpts from Origen, Tertullian, Polycarp, Clement of Alexandria, Justin, Irenaeus, and others – and equally revealing material from their critics, detractors, and persecutors.

Hardcover, 379 pages, £19.16 **£13.41 with subscriber discount**

### Renegade
Martin Luther, the Graphic Biography
*Andrea Grosso Ciponte and Dacia Palmerino*

Five hundred years ago Martin Luther confronted the most powerful institutions of his day, sparking the Protestant Reformation that marked one of the great turning points in history. His story comes vividly to life in this graphic novel.

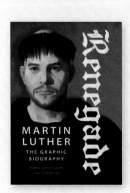

'An earnest take on Luther's life, wonderfully rendered through paintings and collages that dip into the biography at critical moments. YA and adult readers alike will find this work intriguing and informative.'
—*Library Journal,* **starred review**

Softcover, 160 pages, £14.99 **£10.49 with subscriber discount**

# Children's Classics

## Come Again, Pelican

*Don Freeman*

From the creator of *Corduroy*, a newly restored classic picture book that celebrates a child's bond with the natural world. Every summer Ty's family came to camp in their trailer at the same beautiful spot on the white sand dunes by the ocean. And every year, as long as Ty could remember, the same old pelican had welcomed them. This year, as soon as the trailer was parked, Ty pulled on his shiny red wading boots and ran with his fishing pole to look for his friend.

Hardcover picture book, 44 pages, ~~£13.99~~ **£9.79 with subscriber discount**

## Harvey's Hideout

*Russell Hoban and Lillian Hoban*

For big sisters and little brothers in dens, burrows, and houses everywhere. From the author of *Bread and Jam for Frances* comes another children's classic.

'The illustrations amount to Frances in full colour... The Hobans, as usual, know what makes kids kick. Lots of them will find *Harvey's Hideout*.'
—***Kirkus Reviews***

Hardcover picture book, 42 pages, ~~£15.43~~ **£10.80 with subscriber discount**

## If My Moon Was Your Sun

*Andreas Steinhöfel*

A touching story about dementia and the special relationship between grandparents and grandchildren, with an audiobook featuring classical music for children by Georges Bizet and Sergei Prokofiev.

'Steinhöfel tenderly captures a child's fear and understanding of a loved one with dementia... With its loving portrayal of aging, caring for the elderly, and the keen nature of kids' sensibilities, this is a must-purchase for all libraries serving children.'
—***School Library Journal***, **starred review**

Hardcover, 80 pages, ~~£13.99~~ **£9.79 with subscriber discount**

# Peter Waldo

*Several centuries before Luther, a reformer and his band of itinerant preachers rattled the church.*

## CORETTA THOMSON

ON THE FEAST OF THE ASSUMPTION in 1174, a cloth merchant named Peter Waldo stood in the market square of Lyon handing out the last of his money to the poor. 'No one can serve two masters, God and mammon!' he cried (Matt 6:24). 'Citizens and friends, I am not mad, as you imagine... I am urged to this for my own good and yours; for myself, that if hereafter anyone should see me with money, he may say that I have gone mad; for you also, that you may learn to put your trust in God and not in riches.'

Tradition recounts that Waldo had stood there week after week giving out food to famine-ravaged townspeople. Before this, he had provided for his wife and two daughters and commissioned vernacular translations of the New Testament and other texts by Church Fathers. His conversion happened after a companion died of a seizure during a banquet. 'If death had taken me, what would now be my destiny?' Waldo realised with a shock. A few weeks later, a passing troubadour sang of Saint Alexis, who had abandoned wealth, status and family for a life of itinerant poverty. Deeply moved, Waldo invited the minstrel home to hear the story again. The following day he asked a priest which way to heaven was the most perfect. 'If you would be perfect, go, sell what you possess and give to the poor,' was the reply (Matt 19:21).

The real-life conversion of Waldo may not have played out exactly as this legend suggests. In fact, the name 'Peter' does not appear in the extant Waldensian writings until 150 years after his death. What is clear from 12th-century records, however, is that the man now known as Peter Waldo – also called Valdés, or Valdesius in Latin – left his wealth sometime in the early 1170s and began preaching publicly. He exhorted everyone he met to take the scriptures seriously and actually do what Jesus instructed. Everyone, not just clergy and consecrated people, can put Jesus' teachings into practice in daily life. Some joined him and this loosely associated band of mendicant, itinerant preachers was named 'The Poor.' Detractors called them 'sandalled ones' or simply referred to them as Waldensians. As a contemporary, Walter Map, observed in 1179, 'These people have no settled dwellings, but go around two by two, barefooted and dressed

*Coretta Thomson is an editor at* Plough *and oversees its Spanish-language publications. She lives at Fox Hill, a Bruderhof community in upstate New York.*

**Petrus Waldus**
(aus Lyon; lebte ums Jahr 1160; aus dem Lutherdenkmal von Rietſchel).

This 19th-century woodcut engraving of Peter Waldo is based on
a sculpture by Ernst Rietschel (1804–61).

in wool tunics. They own nothing, sharing everything in common, after the manner of the apostles. Naked, they follow a naked Christ.'

Although sometimes thought of as a 'proto-Protestant,' Waldo sought to reform the Catholic Church, not abandon it. An 1180 document believed to have been signed by Waldo declares belief in orthodox Catholic tenets. Private Bible reading in the vernacular was not necessarily forbidden in late medieval France, where literacy was rising in the growing towns. Others before Waldo had left their wealth and within a few decades, Saint Francis of Assisi and Saint Dominic would do the same. Rather than start monasteries, the first Waldensians remained itinerant. They refused to perform any work but preaching lest they be tempted to accumulate wealth. Moreover, they took the radical step of publicly reading, preaching and interpreting scripture as laypeople.

At first the Waldensians were tolerated for their doctrinal orthodoxy. When Waldo and some companions appeared at the 1179 Third Lateran Council and presented Pope Sylvester with a copy of their Bible, the pontiff embraced Waldo, commended their vow of mendicant poverty and allowed them to preach so long as they received permission from their local bishop. Relations quickly soured, however. Some Waldensians became anticlerical and others failed to secure permission to preach. Moreover, women preachers joined the movement in 1180. When Archbishop of Lyon Jean de Bellesmains summoned Waldo and forbade further preaching, he replied, 'We shall obey God rather than men'. In 1184, the Waldensians were condemned as schismatic.

The network grew, with Waldensians appearing in Britain, Germany and Spain. They adopted beliefs considered heretical at the time. Some rejected taking oaths or supporting the death penalty. A few proposed that it was acceptable to confess to a layperson if the local priest was corrupt. Others organised simple Lord's Suppers administered by laypeople. The movement grew too radical for even its founder. In Lombardy, a group of Waldensians settled in communities, rebaptised those who wanted to join them and stated that only those entirely divested of wealth would enter heaven. Waldo expelled them from The Poor in 1205. He died about a year later.

As external dangers mounted, their preaching went underground. In the 1230s, the Inquisition initiated a full-scale persecution against the Waldensians. Still, they spread throughout continental Europe. When towns became too dangerous in the 1300s, The Poor fled to the countryside. In the 1400s, they drew close to the Hussites in Moravia. Crusaders invading the Alpine valleys slaughtered hundreds, causing some Waldensians to abandon pacifism.

When Luther sparked the Protestant Reformation, the Waldensian movement was already over three hundred years old. There was contact between the two groups as early as 1523 and the Waldensians would eventually adopt beliefs and structures of the Reformed Church. Some Waldensians saw this move as betrayal: a loosely structured, persecuted and unaligned movement becoming tied to an established church and a systematic theology. Others have maintained that the measure was the only path to survival.

This year Waldensians in Italy, Uruguay and Argentina – the countries where significant Waldensian communities still exist – celebrate 850 years of history, most of it spent as persecuted minority congregations. A Waldensian emblem depicts a candle and a book with the motto *lux lucet in tenebris* (light shines in the darkness). Waldo's call to let God's word illuminate our lives, to live according to it and to share it with others has not been extinguished. ✦